Graham Brown • Karon Hepner

The Waiter's Handbook

4TH EDITION

Pearson Australia
707 Collins Street
Melbourne VIC 3008
www.pearson.com.au

Acquisitions Editor: Lisa Railey
Project Editor: Bernadette Chang
Production Coordinator: Barbara Honor
Copy Editor: Robyn Flemming
Cover design by Christabella Designs
Typeset by Midland Typesetters, Australia

Printed and bound in Australia by Pegasus Media & Logistics

National Library of Australia
Cataloguing-in-Publication Data

Author: Brown, Graham
Title: The waiter's handbook / authors, Graham Brown; Karon Hepner.
Edition: 4th ed.
Publisher: Frenchs Forest, N.S.W. : Pearson Education Australia, 2008.
ISBN: 9780733993473 (pbk.)
Notes: Includes index.
Subjects: Waiters—Handbooks, manuals, etc.
               Waitresses—Handbooks, manuals, etc.
               Table service—Handbooks, manuals, etc.
Other Authors/Contributors: Hepner, Karon.
Dewey Number: 642.6

Pearson Australia Group Pty Ltd   ABN 40 004 245 943

Foreword

I am both pleased and honoured to have been asked to write the foreword for the fourth edition of *The Waiter's Handbook*.

Reflecting on over 30 years' experience in the hospitality industry, I recall the days when waiting was a trade and the industry offered a waiting apprenticeship. Daily waiting duties included the polishing of fine silverware, and most food was served in the fork and spoon fashion. These were the days when customers were greeted by name ('Good evening, Mr Smith') and were offered their favourite drink on arrival ('May I get you your usual dry martini, Mr Smith?').

Over the years, the food and beverage industry has continued to flourish. Service has become more relaxed; fine dining may be a once-a-year experience, with Australians enjoying casual dining on a regular basis. The expectations of consumers have continued to increase. Today, more and more people go out for a luncheon or dinner with friends or business associates, instead of hosting dinner parties at home.

The hospitality industry in Australia is world class. *The Waiter's Handbook* provides industry professionals and hospitality students with a comprehensive and detailed overview of the skills required for a successful business or career in hospitality. Set out in a user-friendly format, *The Waiter's Handbook* provides the basis for quality customer service and the skill development that is necessary in our industry today.

The hospitality industry is a growth industry in Australia and irrespective of whether a business offers fine dining, a bistro, bar or a club operation, coffee shop, café, function centre or event catering, the customer service requirements remain the same. Employers rely upon skilled employees to act as ambassadors for their business, as they are at the front line of communications with customers. Employers need to be able to have confidence in their employees.

The hospitality industry continues to offer casual employment to students, backpackers and similar persons, and these casual employees complement the full-time and part-time workforce, which generally has had some formal industry training. Most casual employees have had little or no formal training for the hospitality industry, and they attain the majority of their skills in the workplace and on the job.

The Waiter's Handbook is an ideal reference book for hospitality students, as well as for hospitality employers who are seeking to achieve and sustain quality customer service within their business. The publication is a professional training resource and is commended to all students and employers.

May I congratulate the authors on this comprehensive guide to food and beverage service and acknowledge the commitment of the authors to the hospitality industry.

John Sweetman AM

Chairman, Service Skills Victoria

Vice Chairman, Service Skills Australia

Waiting as a profession

The hospitality industry is a range of vastly diverse hospitality businesses, all requiring staff capable of delivering forms of service that need a variety of applied skills. Any one customer can require and experience many different service styles over a single day's journey while eating out from breakfast to supper. In turn, any single waiter or service deliverer can be employed in more than one style of establishment or operation in a working day or working week.

Any waiter who is employed in Australia these days has to try to understand and know it all. Once acquired, this knowledge, and the skills needed to apply it, will equip practitioners with the confidence to make decisions and choices of service delivery that are intelligent, efficient and informed, decisions that will provide outstanding service appropriate to the moment, the occasion and the establishment.

The Waiter's Handbook provides comprehensive instructions on the principles and skills that waiters must master. Only when they have mastered these principles and skills are waiters free, occasionally, to break the rules intelligently and strategically. By 'breaking the rules' I mean making decisions, where necessary and appropriate, to change or challenge standard practice. When this can be done with confidence, serving people as a waiter transcends servitude and ritual and becomes the empowering, creative and satisfying occupation that I have personally known it to be.

I have the highest respect for the work of my industry colleagues, Graham Brown and Karon Hepner, and commend them for nurturing this publication with the appropriate changes and additions to sustain its relevance. The hospitality industry is enriched by their contribution as educators and leaders.

Dur-é Dara OAM
Waiter, Restaurateur and Musician
Former President, Restaurant and Catering Association of Victoria

Contents

Preface

The Waiter's Handbook has been designed as a basic training aid for waiters involved in the service of food and beverages.

Regardless of the styles of the establishments where waiters are employed, the skills of our industry do not differ in their basic application, but only in their interpretation. We believe that good professional service is as important in the café and the informal bistro as it is in the fine dining-room.

The term 'waiter' indicates a professional waiting person regardless of their sex or place of employment. The term 'professional' relates to a waiter who demonstrates both a positive hospitable attitude and the application of technical ability.

As a training manual *The Waiter's Handbook* was designed with both the beginner and the professional in mind. The book is designed to make clear the key objectives of professional food and beverage service and the essential techniques involved in their application.

The main text introduces the reader to all the equipment a waiter is likely to use and its application. There is also a chapter on beverage product knowledge and a new section on dietary and cultural awareness. Food product knowledge, menu terminology and industry jargon are all covered in the extensive glossary. The glossary covers the terms and items most commonly found in today's menus which draw on so many different culinary traditions, as well as explaining the long-established terms of classical French cuisine.

Techniques used throughout the book are described for a right-handed waiter. A left-handed waiter should reverse the techniques, except that all beverages are placed to the guest's right, and both left- and right-handed waiters work in the same direction around the table.

Our book suggests that both food and beverage service should take place clockwise round the table, starting with the guest on the left of the host, and finishing with the host. Establishments that recognise the custom of placing the principal guest immediately to the host's right will often choose to serve in an anti-clockwise direction to accommodate the flow of service around the table. However, in this and in many other points of detail, there is room for differences of practice; the important thing is that there should be consistency within each operation.

Our aim has been to instil confidence in trainee waiters, and to provide the trained and more experienced with a reference work for use throughout their working careers.

Graham Brown

Karon Hepner

Acknowledgements

This book is the result of the hard work and generosity of a great many people and institutions.

We would like to thank Didasko Group (Multimedia Learning Resources) for their strong support and encouragement while this edition of *The Waiter's Handbook* was being written.

We would also like to thank the Coffee Academy, William Angliss Institute of TAFE for contributing coffee illustrations; Matteo's Ristorante for menu samples and for allowing us the use of their premises while this edition was being prepared.

Among the many helpful commentators on the typescript we must particularly thank Dur-é Dara, well known Restaurateur, for contributing her introductory piece 'Waiting as a Profession' and Mark Kirkham, Executive Assistant Manager Harmon Group of Hotels, for drafting the chapter on Room Service.

We would like to thank Jill Adams, Coffee Academy, William Angliss Institute of TAFE, Elizabeth Grist, Didasko Learning Resources and Dino Kapetanakis, Holmesglen Institute of TAFE for their contribution and support in the writing of the Espresso Coffee chapter. We also acknowledge Service Skills Australia for their contribution in supplying information relating to the new Service Industries Training Package 2008 and Didasko Learning Resources for sharing the content relating to dietary and cultural requirements in Chapter 6.

Our appreciation goes to John Sweetman AM, Chairman, Service Skills Victoria, Vice Chairman, Service Skills Australia and the late Ian Ross, Managing Editor of Hospitality Foodservice, for giving their support to training, and encouragement of the development and use of *The Waiter's Handbook* and also to John for contributing the Foreword to the fourth edition. The Australian Institute of Hospitality Management has also supported us and selected *The Waiter's Handbook* as its official handbook.

Domaine Chandon of the Yarra Valley (the Australian winery of Moët et Chandon of Epernay, France) and the wineries of the Goulburn Valley provided the wines used to illustrate this book. Red-Cat Systems (Hospitality Software) of the ACT have very kindly allowed us to base the illustration of point-of-sale billing systems in Chapter 6 on a diagram illustrating their Touch Screen POS System. Thanks also to Flower Flower creative florist of North Fitzroy for their support.

We would like to express our appreciation to our professional industry and educational colleagues who made up the working parties that defined the changes, revisions and additions that needed to be made for this new edition of *The Waiter's Handbook*, particularly Liz Grist and Matt Pignatella.

Many of our professional colleagues, former students and friends gave their time to demonstrate techniques or to be used as models, suffering, without complaint, the long delays and the loss of their free time that painstaking professional photography involves. They included Tim Scott, Cherie Braude, Pamela Habjan, Wesley Orwin, Alison Fernandez, Luke Lenard, George Foundas, Leonie Vanderheyden and Brook Finlay.

The industry professionals who appear in the photograph on the front cover are: Tim Scott—Grand Hyatt Melbourne; Cherie Braude—Grand Hyatt Melbourne; Alicia Lor—Holmesglen Institute of TAFE; and Charles Carrier—Matteo's Ristorante.

Without the support of our partners Rosemary Brown and Nigel Gaudie and families, and the help of these and many other generous helpers and advisers, *The Waiter's Handbook* would have been a very different book. We are extremely grateful to them all.

Graham Brown

Karon Hepner

Chapter 1

Hospitality and the waiter

Why should waiters who work in coffee shops or serve counter meals in pubs feel any less professional in their role than waiters who offer silver service?

Guests have a right to professional service no matter how little they are spending.

When you have completed this chapter you will have a basic understanding of:

❖ Different types of food and beverage operations

❖ The role of the waiter

❖ Career pathways in food and beverage

❖ The organisation of food and beverage service staff

❖ The need for high standards of personal presentation and hygiene

❖ What to do in case of fire

❖ What to do if there is an accident

❖ How to maintain security in a hospitality establishment.

Waiters are employed in a huge variety of establishments in the hospitality industry, but, whatever type of restaurant or other venue they work in, the waiter's basic role doesn't differ.

The number of food and beverage service staff, and their positions in an establishment's hierarchy, depend on the size of the operation and the services it offers. However, the function of the waiting staff, large or small, remains the same, as does the need for the waiters to be professional in what they do.

In any establishment, large or small, formal or informal, the industry demands that the waiting staff should have a professional attitude to their jobs.

Food and beverage outlets

The hospitality industry offers employment to people of differing personalities, backgrounds and skills through a wide diversity of outlets serving food and beverages.

Hospitality may be defined as meeting the needs of guests in a variety of establishments, including all the following, at which service staff or waiters are employed:

❖ *Coffee shops,* which offer coffee, snacks and often light meals through to supper items. Coffee shops require fast service to ensure a fast turnover of customers.

❖ *Food halls/food courts* have taken over from traditional cafeterias. They offer a wide variety of foods, which guests select for themselves. Service staff are responsible for clearing the eating areas.

❖ *Bistros/pubs (counter meals).* This style of service applies in casual or pub environments. The food, usually main meals, is either collected by the guests from a counter or served by service staff. An essential element of this type of meal service is speed.

❖ *Casual dining restaurants (cafés/bistros).* Here, appearance and atmosphere provide an environment for casual dining, but table service is offered. Informal restaurants don't always have a licence to serve alcohol, but they may have BYO facilities. Service staff must be capable of friendly informality in their dealings with guests while remaining professionally efficient.

❖ *National or 'ethnic' restaurants.* In these popular establishments, the style of the service may be as much a part of the cultural experience offered to guests as the food itself.

❖ *Functions (receptions/banquets/conventions).* In these cases, the number of guests and the style of function vary enormously, so they demand extreme flexibility from both management and service staff.

❖ *Fine dining restaurants* usually have a suitably comfortable or impressive ambience for the fine cuisine on offer. Service staff are expected not only to be discreetly professional in what they do, but also to be highly skilled and knowledgeable.

The role of the waiter

The term 'waiter' includes food and beverage service personnel of either sex.

It is the role of the professional waiter to ensure that guests enjoy a satisfactory total dining experience; the job involves much more than simply serving food or beverages.

To fulfil this role adequately, a waiter needs a range of qualities and attributes, including a pleasant personality, honesty, efficiency and punctuality. Also, a waiter must always be fastidious about self-presentation and personal hygiene.

A professional waiter will also have a good knowledge of the products being served, what they consist of and how they are presented. Good waiters will also understand the organisation of the establishment in which they are working, and how other members of the team contribute to the dining experience of the guests.

THE AUSTRALIAN WAITER

While the role of the waiter is essentially the same everywhere—to ensure that guests have a satisfactory dining experience—it requires differences in approach in different circumstances. In Australia, where people are in general less formal in their dealings with each other than in older societies such as those of Europe or Japan, the strict formality of the traditional European waiter is often not appropriate. This doesn't mean that Australian waiters should be any less attentive to the needs of their guests, but simply that they can and should project the self-confident non-servile individuality that is a most attractive aspect of the Australian character.

Australian waiters should not try to imitate other service cultures, as good Australian service concentrates on the specific needs of the guests—service without servility.

DUTIES OF THE WAITER

The work of the waiter includes:

❖ preparation and maintenance of the work area

❖ maintaining good customer and staff relations

❖ making recommendations and assisting guests in making selections

❖ order taking and recording

❖ service and clearing of food and beverages.

CUSTOMER COMPLAINTS

In every business, there will be times when people complain about some aspect of the service or produce. When a customer complains, it is the waiter's job to make sure that the complaint is dealt with in such a way that the customer will remain a customer in the future. It is far easier to keep an existing customer than to find a new one.

Many dissatisfied customers don't complain, but simply don't come back. While they may not tell you, they will probably tell their friends about whatever it was that upset them. People who *do* complain are therefore doing the business a favour. They are giving you the chance to retain their custom by handling their complaint politely and professionally. They are also giving the establishment an opportunity to improve its service or product in future.

When a customer complains, you must listen to the complaint carefully and deal with it respectfully. When you have established exactly what the complaint is, apologise and try to find a solution. Act upon the complaint immediately. Don't blame anyone else and don't take the complaint personally. If you allow your personal feelings to get in the way, you are at risk of losing your composure and making the situation worse.

Career pathways in food and beverage

In Australia you can build a career in either beverage or food, or a combination of both. The table on page 5 illustrates a number of career pathways for people with a Hospitality qualification. The positions listed may be found in a variety of establishments, including cafés, cafeterias, fast food outlets, restaurants (licensed and unlicensed), nightclubs, retail liquor outlets, function and convention centres, large- and small-scale caterers, casinos, resorts, registered clubs, motels, hotels, international hotels, guesthouses, executive apartments, caravan parks, hospitals, retirement and nursing homes, the defence forces, airlines, railways and cruise liners.

To progress up the career ladder, you will be expected to undertake specific accredited training programs that align with the national training package.

There are various pathways to success.

1. You can complete Certificates I and II in Hospitality at school and/or at a TAFE or private training provider to provide you with the basic skills, and then find a job as a food and beverage attendant. You can continue your education as you go, by being assessed in the workplace through a TAFE that provides a Skills Recognition Audit program. This provides employees and employers the benefit of industry recognition for what has been learned at work, in unfinished courses, through life experience and in other industries. The program assesses candidates against the national competency standards in such areas as food and beverage, gaming, guest services, kitchen attending and commercial cookery.

2. Or you can undertake a traineeship or apprenticeship and complete your first certificate on the job. You can continue your education as you work, learning and getting paid.

3. Or you can complete a diploma or advanced diploma at a TAFE or private training

Hospitality and events qualifications and job roles

Hospitality qualifications and job roles

SIT10207 Certificate I in Hospitality
- bar useful
- coffee shop assistant
- waiter
- food and beverage runner
- housekeeping assistant
- porter

SIT10307 Certificate I in Hospitality (Kitchen Operations)
- kitchen attendant
- larder hand
- sandwich hand

SIT20207 Certificate II in Hospitality
- bar attendant
- bottle shop attendant
- catering assistant
- food and beverage attendant
- housekeeping attendant
- porter
- receptionist or front office assistant
- gaming attendant

SIT20307 Certificate II in Hospitality (Kitchen Operations)
- breakfast cook
- short order cook
- fast food cook

SIT20407 Certificate II in Hospitality (Asian Cookery)
- short order cook
- cook

SIT30707 Certificate III in Hospitality
- bar attendant
- barista
- wine waiter
- front desk receptionist
- housekeeper
- gaming attendant

SIT30807 Certificate III in Hospitality (Commercial Cookery)
- cook

SIT30907 Certificate III in Hospitality (Asian cookery)
- cook

SIT31007 Certificate III in Hospitality (Catering Operations)
- cook
- leading hand or food service assistant

SIT31107 Certificate III in Hospitality (Patisserie)
- patissier

SIT40307 Certificate IV in Hospitality
- food and beverage supervisor
- front office supervisor
- concierge
- butler
- gaming supervisor

SIT40407 Certificate IV in Hospitality (Commercial Cookery)
- chef
- chef de partie

SIT40507 Certificate IV in Hospitality (Asian Cookery)
- chef
- chef de partie

SIT40607 Certificate IV in Hospitality (Catering Operations)
- catering supervisor
- caterer

SIT40707 Certificate IV in Hospitality (Patisserie)
- chef patissier
- chef de partie

SIT50307 Diploma of Hospitality
- restaurant manager
- kitchen manager
- front office manager
- housekeeper
- sous chef
- gaming manager
- motel manager
- unit manager (catering operations)

SIT60307 Advanced Diploma of Hospitality
- food and beverage manager
- area manager or operations manager
- rooms division manager
- executive housekeeper
- secretary manager
- executive chef
- cafe owner or manager
- motel owner or manager

Events qualifications and job roles

SIT30607 Certificate III in Events
- event assistant
- event administrative assistant
- event operations assistant
- event operative
- conference assistant
- exhibitions assistant

SIT50207 Diploma of Events
- event coordinator
- venue coordinator
- conference coordinator
- exhibitions coordinator

SIT60207 Advanced Diploma of Events
- event manager
- venue manager
- conference manager
- exhibitions manager

Source: Service Skills Australia, 2007. Service Skills Australia is the national Industry Skills Council for the service industries, and is one of 10 independent, industry-led Industry Skills Councils funded by the Australian Government Department of Education, Employment and Workplace Relations (DEEWR) to support skills development for Australian industry.

provider, or maybe even a degree course at university, working part time as a food and beverage attendant to gain some work experience.

4. If you are to progress to a management position, you will be expected to complete a tertiary course and to have had extensive experience in the industry. That means long hours of work in a very demanding environment.

The organisation of food and beverage service staff

While jobs advertised may refer to the formal classifications of an industrial Award, those classifications don't detail the actual duties of each job. These will vary from establishment to establishment, according to its size, the nature of its business and the traditions of the organisation. Positions commonly found include:

❖ *Food and Beverage Manager.* In larger operations, a Food and Beverage Manager is usually responsible for the success of the food and beverage operations from a business point of view. He or she will be responsible for such matters as compiling the menus (in consultation with the kitchen) to make sure that the required profit margins are achieved, purchasing food and beverage items, and staff recruitment and training.

❖ *Restaurant Manager.* In operations where there are several bars and restaurants, such as a large hotel, each restaurant may have its own manager responsible to the Food and Beverage Manager. He or she will be responsible for the work of the staff within that restaurant and for seeing that the policies of the Food and Beverage Manager are carried out. Either the Restaurant Manager or the Head Waiter will be responsible for staff duty rosters.

❖ *Head Waiter/Supervisor.* The Head Waiter, or Supervisor, is responsible for all the service staff in the restaurant and for seeing that all the preparation, service and clearing work is carried out efficiently. In smaller establishments, he or she may also be responsible for taking reservations and for greeting and seating guests. In larger establishments, there may be a special Reception Head Waiter with these duties.

❖ *Station (Head) Waiter.* The Station Head Waiter, or Captain, is responsible for the service of a **station**, or group of tables. He or she takes the orders and carries out the service at the tables of the station, assisted in larger establishments by less experienced and less knowledgeable staff. Each station may have its own workstation, or **sideboard**. (This sideboard is also called the waiter's 'station'.)

❖ *Waiter.* If the stations are looked after by a service team, less experienced waiters are responsible to the Station Head Waiter. They perform duties such as plate service of dishes and the service of sauces, sometimes assisted in the simplest tasks by a trainee.

A trainee waiter is sometimes called a **commis** (pronounced *commie*) waiter. He or she should not be confused with a commis cook—that is, a cook in training, working under the chef.

These days, the job of waiters is often to take responsibility for the total experience of the guests at their tables, from arrival to departure. The job of individual waiters is therefore even more important than if they were part of a service team.

❖ *Wine Waiter (Sommelier).* The wine waiter is responsible for the service of all alcoholic drinks to the tables. He or she must, of course, have a thorough knowledge of the wines on the establishment's wine list, and be able to recommend suitable wines to accompany the various menu items—and, of course, know how to serve them correctly.

Sommelier (pronounced *som-may-lee-ay*) is another word for a wine waiter.

Presentation and hygiene

One all-important aspect of the professionalism of waiting staff is the attention they give to personal hygiene and presentation. When at work, in or out of uniform, a waiter must invariably be absolutely clean and tidy in all respects. This is the first and most obvious sign that waiters are professional in their approach to their work.

The first (visual) impression of waiting staff received by guests comes from the waiters' appearance. First impressions are extremely important for the commercial success of an establishment.

Good grooming and meticulous attention to personal hygiene not only express a positive attitude towards guests but also build self-confidence in the individual waiter.

For health reasons, high standards of personal hygiene are essential for all workers involved in the service of food and beverages.

High standards of hygiene in the waiting staff are also essential if guests are to enjoy their dining experience. Their enjoyment will be considerably lessened if their food or drinks are served by a waiter with bad breath or dirty fingernails.

Waiters must select their footwear with care and pay attention to good posture when standing and moving. It is almost impossible for a waiter with sore feet or an aching back to maintain a pleasant and helpful attitude to guests.

Waiters' hairstyles should be suitable to the establishment in which they are working. In general, they should be in tune with current fashions, but hair must be tidy and should be swept away from the face. This projects an air of self-confidence and is also more hygienic.

Unobtrusive jewellery, make-up and perfume may all be used, but with discretion. They must not be overdone.

Whether the waiter's uniform is provided by the establishment or not, it is the waiter's responsibility to ensure that the uniform is kept clean and well-presented at all times.

When uniforms are being selected, it is important that their size and design make adequate allowance for the extensive body movement a waiter's work demands. Natural fibres are preferable to synthetics because they allow the body to breathe and they are safer in case of burns.

HYGIENE CHECKLIST

In law, hygiene is your personal responsibility. You could be prosecuted if a customer suffers through your failure to maintain good standards of hygiene.

❖ Keep your uniform clean and well pressed.

❖ Wear comfortable shoes and keep them clean.

❖ Never let any hair fall into food. Adopt a style that is easy to keep tidy and will keep your hair off your face.

❖ Shower every day and wear a deodorant to eliminate body odour.

❖ Jewellery and perfume should only be worn in accordance with the establishment's rules and state hygiene and food regulations.

❖ Wash your hands thoroughly every time you go to the toilet.

❖ Don't smoke anywhere near food. If you do smoke, wash your hands afterwards.

❖ Keep your hands away from your face, especially your mouth and nose.

❖ Cover any cuts or burns to avoid the risk of contaminating food.

❖ Report any illness or infection to your supervisor.

❖ All food service attendants must have completed an AHRP-accredited occupational food hygiene program.

Fire safety

One of the first things to learn when you start a new job is the establishment's safety code for the staff and guests. You should always follow the rules of your establishment, but there are a few universal rules that must be followed everywhere.

❖ Raise the alarm immediately you discover a fire and warn people around you to move away.

❖ Don't panic. The premises should be evacuated in a quiet and orderly manner.

❖ Never put yourself or others in danger by staying near a fire you cannot control.

❖ Use any fire-fighting equipment only in accordance with the establishment's rules.

❖ Follow any safety and emergency procedures.

❖ Don't return to the scene of a fire until given the all clear.

❖ Assemble at the stated fire point.

Accidents (staff or guests)

Remember that you will be held responsible in law for any accident that occurs because of your negligence. Health and safety legislation has been laid down to prevent and deal with accidents.

The various state Health and Safety at Work Acts are concerned with protecting the health and safety of employees and the public. It is a legal obligation to ensure, as far as reasonably practicable, the safety, health and welfare of employees. Supervision and training must be given to all staff. It is the responsibility of employees to ensure that their acts or omissions don't affect others adversely.

In the event of an accident, staff should:

❖ notify a First Aider immediately

❖ comfort and reassure the injured person

❖ contact emergency services, following establishment procedures

❖ record the accident in accordance with the establishment's method of documentation.

Work safety and the law

It is a legal requirement that the working environment should be safe. Any hazards or potential hazards that threaten the safety of staff or guests should be identified and preventive action taken, following establishment procedures. These procedures should have been drawn up in accordance with health and safety regulations. Hazards or potential hazards should be reported as soon as possible to lessen the risk to staff or guests.

Quick preventive action could avert injury to staff or guests and avoid damage to equipment and buildings. By doing this, you are complying with the law.

Security

It is the duty of all staff to help maintain security in the following areas:

❖ work areas

❖ staff facilities

❖ client facilities

❖ public areas

❖ storerooms

❖ cellars

❖ refrigerators

❖ freezers.

Keys, property and the areas listed above should be secured against unauthorised access at all times. This will help to guard against damage and pilferage.

Any property that goes missing, whether it belongs to staff or to guests, should be reported to a supervisor at once and appropriate action taken.

Any suspicious persons should be challenged politely or reported to security.

Any lost property should be documented and kept secure until claimed.

QUESTIONS

1. Name eight different sorts of establishments where food and beverage service are required.

2. Describe four attributes of a professional waiter.

3. What are five important duties carried out by a waiter?

4. What forms of training allow you to acquire food and beverage qualifications on the job?

5. What is a commis waiter?

6. Who is responsible for personal hygiene?

7. There are 11 points on the hygiene checklist. What are they?

8. Name two things you should never do at the scene of a fire.

9. How can you keep the work environment safe?

Chapter 2

The menu

Why bother to provide a menu if the descriptions of the menu items are not true to the items served?

When you have completed this chapter you will have a basic understanding of:

❖ The framework of the modern menu

❖ The sequence of the courses

❖ Food items that are not part of the framework of courses

❖ The different types or classes of menu.

The term '**menu**' has at least three meanings for the waiter. 'Menu' means:

❖ the range of food items served in an establishment, including their organisation into a number of courses

❖ the arrangement by which the items are offered (the type of menu, as in 'set menu', 'à la carte menu', etc.), and

❖ the physical object on which the list of these items (and courses) is written for guests to choose from.

This chapter is concerned with the menu in the first two of these senses. The handling of the menu as a physical object and its presentation to guests is dealt with in Chapter 6.

The structure of the menu

There is necessarily a menu of some kind or other for every meal, however simple. For example, cornflakes followed by toast and marmalade is a popular breakfast menu. However, the menu this chapter is concerned with is the menu for a main meal of the day—luncheon or dinner.

The traditional French menu, or '**classic menu**', had more than 12 courses. It offered the diner a wide variety of items served in a well-understood traditional sequence. Modern menus have fewer courses, but they may cause confusion as to the appropriate sequence of service.

Menus are laid out so that the different courses appear in the order in which they would normally be served. They are usually presented in a framework of five courses, as follows:

❖ appetisers

❖ soups

❖ entrées

❖ main courses

❖ desserts.

Appetisers are such items as hors d'oeuvres, pâtés and oysters natural, which are designed to stimulate rather than to satisfy the appetite.

Soups may be thick (potages) or thin (consommés). The less common French word **soupe** (with an *e*), often used in *soupe du jour* (soup of the day), can mean either a thick or thin soup. Soups are usually hot, but can be served chilled—for example, vichyssoise.

In the classic French menu, the **entrée** was a course served between the fish and the main meat course. In the modern menu, the term is used to cover such items as small helpings of pasta dishes, seafood crêpes, elaborate salads, miniature sausages, or fish (if not chosen as the main course). The entrée is a course that certainly does more than stimulate the appetite but is not so substantial as to make the main course unwelcome.

Many Australian establishments list all these three courses under the general heading of 'starters' or sometimes 'entrées'.

The term 'entrée' can cause even more confusion because, sometimes, especially in the United States, it is used to mean the main course itself, although the word literally means 'entrance' or 'way in'—that is, a starter.

There is, fortunately, no possible confusion about what the main course is. It is the most substantial course of the meal. Guests usually choose their main course first and then select other courses to suit it. Similarly, when chefs design menus, they usually start with the principal or main course and then plan the other courses to complement it.

Dessert is another term that can cause confusion. In the modern menu it is used to mean the sweet course at the end of a meal, although the term is occasionally reserved (particularly in Britain) for the fruit and nuts served after the sweet dish or pudding has been cleared.

SEQUENCE OF COURSES

In most formal circumstances the courses are served in the order assumed in the standard modern menu, but this structure may not readily apply to the menus actually offered in many less formal or non-traditional establishments. When this is so, it is the responsibility of the waiter to determine in what order the guests wish their courses to be served.

Items outside the menu structure

Not all the food items served with a meal are included in the formal structure of the menu.

PRE-DINNER FOOD ITEMS

Items served before the first course may include:

❖ canapés

❖ crudités

❖ bite-sized hot or cold hors d'oeuvres.

SAUCES AND ACCOMPANIMENTS

Now that our eating habits draw on so many different cultural traditions, the range of different sauces and accompaniments available is huge. It is appropriate for some of these items to be placed on the table for the guests to help themselves; others should be offered to the diners by the waiter (see Chapter 8).

It is the waiter's responsibility to make sure that his or her workstation's mise-en-place (see Chapter 4) includes the necessary sauces and accompaniments—such as different kinds of mustard, coarse-ground black pepper, chutneys and ketchups—appropriate to the menu of the establishment.

CHEESE

Cheese may be ordered prior to dessert (as is the usual custom in France, for example), in place of dessert or after dessert, depending on the guest's preference. While some guests may prefer cheese to be served while they are still enjoying table wine, others may prefer it to be served later, with coffee and port.

The correct procedure for serving cheese, and the appropriate accompaniments, depends on the types of cheese being served. Fashions change in what accompaniments are considered 'appropriate' and different establishments have different practices. Accompaniments may include such things as dried or fresh fruits, nuts, fruit pastes (for example, quince paste) and salad items. There is no one correct procedure applying to all establishments.

ACCOMPANIMENTS TO COFFEE

Like pre-dinner food items, accompaniments to coffee are not included in the menu framework, but are an addition to it.

The range of items that may be offered with coffee is limited only by one's imagination. The service of after-dinner mints is standard practice, but many establishments now regard their coffee accompaniments as their personal signature. These items may be petits fours, personalised chocolates, biscuits or glacé fruits, for example.

Types of menu

The different types or classes of menu are distinguished by the variations in the selections offered and by their pricing structure. Menu types include à la carte, table d'hôte, set menu, carte du jour, degustation menus and cycle menus.

The 'type' of menu, in the sense used here, shouldn't be confused with the presentation of the menu—whether it is printed, or written on a blackboard, or presented in some other way.

À LA CARTE MENU

An **à la carte** menu offers choices in each course and where each item is individually priced and charged for. Menu items when selected by the guests are cooked to order. The literal meaning of the French words *à la carte* is 'from the card'.

The term 'à la carte' when applied to a restaurant is often misinterpreted. The term doesn't relate to a particular type of establishment, or to its pricing, or to the services it offers; it refers solely to the type of menu and to the fact that the food is cooked to order.

TABLE D'HÔTE

A **table d'hôte** menu offers some (usually limited) choice and is charged at a fixed price per person for the whole menu. *Table d'hôte* is, literally, French for 'the proprietor's (mine host's) table'.

A modestly priced 'business lunch', in which three or four items only are offered in each course and the guest pays a fixed price for the whole meal, would be a typical use of the table d'hôte menu.

More exclusive restaurants also often make use of the table d'hôte menu, as its limited number of menu items allows the chef to select fresh ingredients of the best quality and to treat each dish with maximum attention. Because a more limited range of choice has to be catered for than in a typical à la carte menu, there is less wastage.

Table d'hôte menus are popular for festive occasions such as Christmas Day and Mother's Day.

SET MENU

A **set menu** is one that offers set items (one for each course) prearranged by the host. Set menus are used mainly for functions, such as weddings and banquets.

CARTE DU JOUR

Carte du jour literally means 'card of the day'. A **carte du jour** menu offers choices that are available for a particular day only. It allows the chef to offer a list of 'specials' or variations in addition to a pre-printed à la carte menu, or it can be used as a table d'hôte menu prepared for use on the one day only.

DÉGUSTATION MENU

Dégustation literally means 'tasting'. A **dégustation** menu lists a range of items, usually specialities of the establishment, which are served in small portions.

 *THE UNICORN*

Starters

Seafood and vegetable broth with a garlic and herb bruschetta. 9.00

Thai chicken parcel, miso and pickled ginger dressing. 10.50

Chargrilled baby octopus on a rocket salad with tomato, basil and olive oil. 11.50

Thin slices of rare peppered beef
topped with a light gorgonzola mousse, layered between crisp parmesan sheets. 11.90

Puff pastry case filled with boneless braised oxtail, artichokes and green peppercorns. 10.50

Tender duck confit on a beet leaf and radicchio risotto. 12.00

Thin spaghetti tossed with Tasmanian black mussels, tomato salsa
and a piquant extra virgin olive oil scented with fresh herbs. 10.50

Large potato gnocchi, pan-fried in olive oil, topped with sautéed prawns
and a dollop of basil-flavoured mascarpone cream cheese. 14.00

Risotto of fresh beetroot with kangaroo prosciutto. 10.00

Main Course

Chargrilled eye fillet medallions, served with creamed potatoes and
rocket leaves with a seeded mustard dressing. 25.50

Lamb fillets marinated with chilli and fennel seeds, chargrilled,
served on Mediterranean vegetables tossed with herb butter. 22.50

Loin of veal crumbed with parmigiano and herbs, pan-fried,
served on an eggplant risotto with lemon aïoli. 22.00

Free-range chicken breast, marinated with herbs and olive oil, roasted,
served with wild mushroom ravioli. 21.50

Chargrilled calf's liver on a potato purée with sautéed spinach,
roast capsicum, baby caper and balsamic dressing. 19.90

Saddle of rabbit, rolled with a truffle mousse, baked in pastry,
served on creamed winter vegetables. 20.50

Grilled Tasmanian salmon, spiced tomato fennel compote,
roast almond dressing. 23.50

Side Orders

Rocket salad with parmesan shavings. 6.50

Caesar salad. 8.90

Steamed vegetables with chive butter. 6.50

Warm baby spinach leaves tossed with olive oil, lemon and smoked bacon. 7.00

Chargrilled Mediterranean vegetables tossed with herb butter. 6.50

Chunky potato fries with rosemary and garlic. 4.50

An à la carte menu

GKs

Table d'hôte Menu

$55.00 per person

Starter

Warm asparagus vichyssoise soup

South Australian Ceduna oysters

Sumac-spiced yellowfin tuna sashimi,
Vermicelli rice noodle salad,
Japanese dashi dressing

Mediterranean vegetable terrine
with eggplant, zucchini, red capsicum and chèvre
tomato gazpacho coulis

Main Course

Individual lamb rump, roasted with rosemary and garlic,
with sautéed artichoke hearts and green beans and a warm 'hummus' chick-pea purée

Seared fillet of Queensland wild barramundi
White cannelloni bean cassoulet with steamed black mussels, asparagus spears and curry oil

Ballotine of chicken filled with mushroom mousse,
served with gnocchi in a porcini cream sauce

Dessert

Caramelised pear pizza with mascarpone

Silvan Estate raspberries in a chiboust cream, glazed brulée style
honey-flavoured madeleine sponge-cake biscuits, raspberry coulis

Chilled pistachio-flavoured cream custard,
with pineapple crisp and a coconut sauce with tapioca

Coffee and Chocolates

A table d'hôte menu

Black and White
Annual Dinner

Starter
Tasmanian salmon, sugar-cured 'gravlax' style, with balsamic essence,
seared rare, sautéed kipfler potato salad with wholegrain mustard

Main Course
Boned saddle of Western Australian white rabbit
filled with a sweet roasted garlic mousse,
with a rabbit risotto finished with Thomas Blue gorgonzola-style
cheese

Dessert
Warm Valrhona chocolate, self-saucing, white ganache fondant,
chocolate ice-cream and vanilla bean crème anglaise

Coffee and Petits fours

A set menu

Menu Degustazione

La Nostra Pro posta di Assaggi.
A meal of 'tastes' from our chef allowing you to sample a variety of flavours and cookery techniques.

Trio di Sarde
Filleted Western Australian sardines, three ways...
Grilled with a Napoli chutney, Marinated 'Escabeche' style, and crisp fried with an anchovy aioli.

Agnolotti di Coniglio con Purée di Sedano-rapa di Verona
House-made rabbit ravioli on a celeriac purée with a lemon thyme jus.

Cotolette di Vitello alla Milanese con Risotto di Melanzane
Loin of veal crumbed with Parmigiano and herbs, pan-fried,
served on an eggplant risotto with lemon aioli.

Il Dolce
Warm strawberry and rhubarb shortcake... Milk chocolate terrine with espresso anglaise...
Zabaglione parfait wrapped in dark chocolate leaf... Orange and Campari sorbet.

Caffe Corretto
Espresso corrected with 15ml grappa

$36 per person
$50 per person with each course matched with a glass of wine.

'The general view that Italian cookery is coarse and heavy and a poor relation of French cookery is entirely mistaken as is the impression that it is limited to pasta and rice dishes. Italians are dedicated to the art of good eating and drinking. It is this experience that we wish to share with you.'

Matteo and Franca Pignatelli

Dégustation menu

CYCLE MENU

A cycle menu is a group of menus that are rotated on a set cycle. Cycle menus are usually used in the institutional sector of the industry—for example, in hospitals and prisons, on airlines and in employee food-service operations (works canteens, etc.).

The cycle menu is used to avoid boredom for both customers and staff, and also to ensure that the diet of the people in the institution is sufficiently varied to be healthy. In a hospital, for example, the cycle would be set to fit the average length of stay of the patients.

Menus should not be designed to a seven-day cycle as this results in the same items always being served on the same day of the week, producing a boring predictability.

QUESTIONS

1. What are the five courses most commonly found on a modern menu?

2. Why are hors d'oeuvres always small items?

3. What is the French name for a thin soup?

4. Which course on a menu do guests usually choose first?

5. What items are provided outside the structure of the menu?

6. Describe an à la carte menu.

7. What is a typical use of a table d'hôte menu?

8. What sort of menu is most commonly used for large functions?

9. What does *carte du jour* mean in English?

10. What are the advantages of using a cycle menu in an institution?

Chapter 3

Food service equipment

Why should a guest be embarrassed by having to use inappropriate equipment for a particular dish?

When you have completed this chapter you will have a basic understanding of:

❖ The appearance and use of the various types of cutlery used for the eating and service of food

❖ The various types of crockery and tableware used for the eating and service of food

❖ The various pieces of equipment required for the preparation of food in the dining room.

There is a huge variety of cutlery and tableware in use in hospitality establishments today. Not only does the basic cutlery come in a variety of styles, but the range of specialist equipment is almost endless. This chapter doesn't attempt the almost impossible task of describing every piece of equipment a waiter may encounter; what it does is to introduce all the equipment used by waiters commonly found in modern establishments. Remember, though, that there is a great variety of possible styles so that the illustrations are only representative of examples of particular pieces of equipment; other styles may be equally common and valid.

Glassware and beverage service equipment are dealt with in Chapter 13.

Cutlery

BASIC CUTLERY (SILVERWARE) ITEMS

Large fork (table fork) used as

❖ main fork or joint fork

❖ serving fork (see Chapter 8)

Large knife (table knife) used as

❖ main knife or joint knife

Small fork (dessert fork) used as

❖ entrée fork

❖ pasta fork

❖ salad fork

❖ dessert (or pudding) fork

❖ fruit fork

Small knife used as

❖ side knife (for buttering bread or spreading pâté)

❖ entrée knife

❖ cheese knife

❖ fruit knife

Steak knife
Note the serrated edge

Fish knife used for

❖ fish

❖ serving delicate or large items (see Chapter 8)

Fish fork (webbed fork) used for

❖ fish

Large spoon (tablespoon) used for serving (see Chapter 8)

Medium spoon (dessert or pudding spoon) used for

❖ desserts (see Chapters 4 and 6)

❖ pasta (in some establishments)

Soup spoon used for

❖ soup

❖ pasta (in some establishments)

Small spoon (teaspoon) used for

❖ tea and coffee

❖ cocktails (e.g. prawn or fruit cocktail)

❖ ice cream

❖ dessert coupes

❖ sugar spoon

Demitasse spoon (teaspoon) used for

❖ demitasse (small coffee) cups

OTHER CUTLERY ITEMS SOMETIMES PLACED ON THE TABLE

Parfait spoon used for

❖ desserts

❖ ice cream

Oyster fork

Snail (escargot) fork

Snail (escargot) tongs

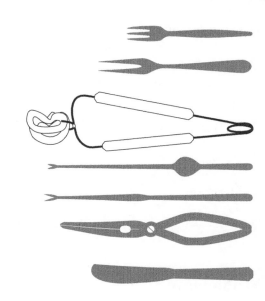

Lobster (crayfish) picks

Lobster (crayfish) cracker

Butter knife used for

❖ serving butter or pâté

Cake fork used for

❖ cakes

❖ pastries

Cheese knife used for

❖ serving cheese

Tea strainer

Sugar tongs

CUTLERY ITEMS USED FOR SERVING

Carving knife

Bread knife

Gâteau (cake) slice

Soup ladle

Nutcracker

Tableware

Common items of tableware (also called crockery or china) include:

Cover plate
Used as a presentation or showplate in the setting. Can be an ordinary large plate also used as a service plate or an underplate for service.

Large plate (dinner plate)
Used for the main course. Oval plates are sometimes used instead of round ones.

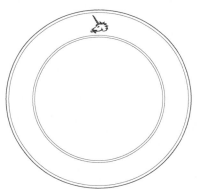

Middle-sized (entrée) **plate**

Used for entrées and also for salads, cheese and fruit. Can be oval instead of round.

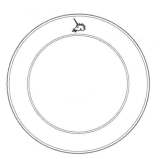

Small (side) **plate**
Used for bread and bread rolls, also for cheese, fruit and cake.

Soup bowl

Used for cream soups, also as an oatmeal (porridge) bowl and for breakfast cereals.

Pasta bowl

Used for all styles of pasta.

Consommé bowl and saucer

Used for clear soups. Note that the consommé bowl is always served on a matching saucer.

Large soup tureen (with lid)

Often used for service from a guéridon rather than on the table (see Chapter 10).

Coupe

A stemmed vessel made of glass, silver or stainless steel. Used for cocktails (seafood or fruit), desserts and ice cream.

Ravière

Oval or rectangular dish, used primarily for pasta or for presenting hors d'oeuvres.

Ramekin

Comes in various sizes. Used for baked eggs, custards and soufflés.

Tea or coffee cup and saucer

Sizes vary considerably. Some establishments use a smaller size for coffee than for tea.

Demitasse and **saucer**/**small coffee cup** and **saucer**
Often used for black or Turkish coffee. (*Demi-tasse* is
French for 'half-cup'.)

Coffee pots
Various styles, including:
Cona pot (Cona is a trade name)

Teapot

Coffee or **tea plunger**

Hot-water pot

Tea infuser

Sugar bowl

Long-spout coffee pot

Milk or **cream jug**

Salad bowl (individual)

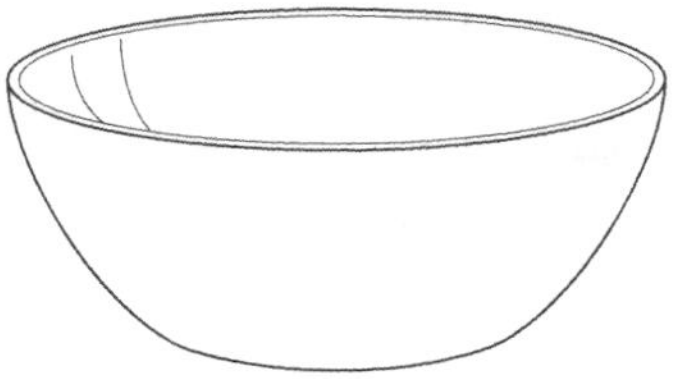

Salad bowl (for table service)

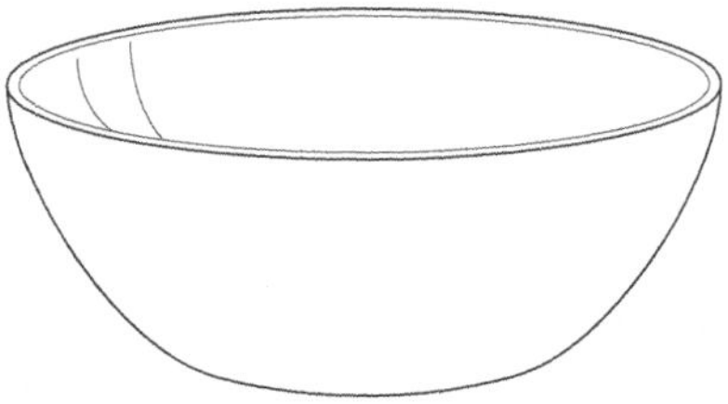

Oyster or **mussel plate**

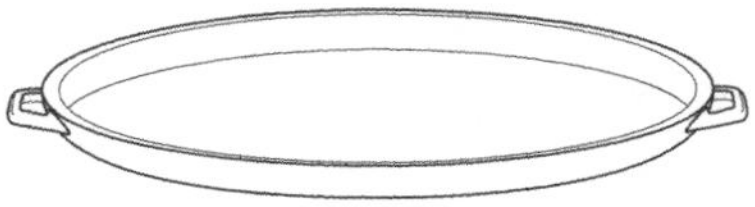

Snail (escargot) **plate**

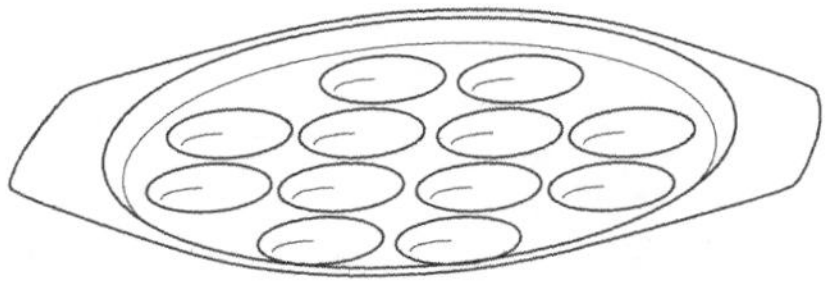

Sauce-boat (saucière)

Salt and pepper set
Often called 'the **cruet**', a term that can include a mustard pot also. The salt pot has a single hole so that salt can be poured neatly at the side of the plate. The pepper pot has three or more holes so that the pepper can be shaken evenly over the dish.

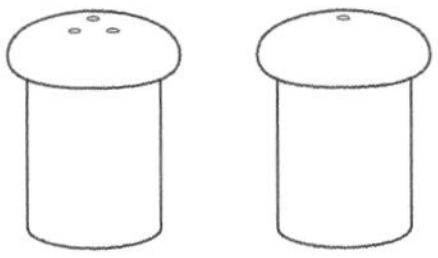

Pepper mill (grinder)

Plate cover (cloche)
There are two styles:
Flat plate cover

Dome

Butter dish

Butter pad

Bud vase

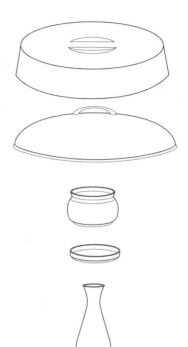

Large equipment

Warming racks (guéridon service)

Guéridon trolley

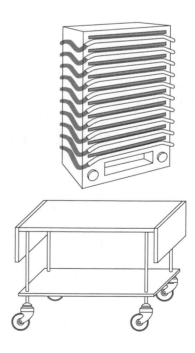

Flambé trolley

Carving trolley (dome) (see Chapter 12)

Flambé pan/Chafing dish (see Chapter 12)

Réchaud (cooking lamp or warmer)
(see Chapter 12)

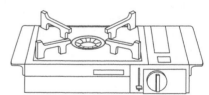

Gas cylinder burner

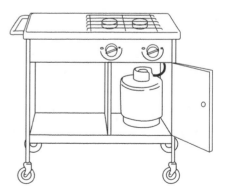

QUESTIONS

1. What are the uses of a small fork?

2. What sort of edge does a steak knife have?

3. What can a teaspoon be used for apart from stirring tea?

4. What do snail tongs and snail fork look like?

5. What is a cover plate used for?

6. What is a ravière?

7. Describe two sorts of coffee pots.

8. What two sorts of plate cover are there?

9. What is a réchaud?

Chapter 4

Food service preparation

Who suffers when guests call you over and complain that their glass is chipped, or that there is lipstick on the edge of the glass, or that there is no salt in the shaker?

Guests are inconvenienced and their dining experience may be spoiled, but it's the waiter who suffers!

When you have completed this chapter you will have a basic understanding of:

❖ Taking reservations

❖ The use of floor plans

❖ The need for an appropriate atmosphere

❖ Clothing tables

❖ Preparation of service stations

❖ The laying of covers

❖ A variety of napkin folds.

A guest's first impression can make or break the total dining experience. Careful and thorough preparation of the dining area before guests arrive is therefore essential. While standards are set by management, it is the responsibility of all employees to maintain those standards by demonstrating individual professionalism in their work.

Taking reservations

Before you take a reservation, make sure you know the answers to the questions you are likely to be asked—for example:

❖ What kind of cuisine do you offer? (French, Italian, Cantonese, modern Australian, etc.)

❖ What style of menu do you offer? (à la carte or table d'hôte?)

❖ Do you accept credit cards? Which ones?

❖ Are you licensed or BYO? Or both? If BYO, is there a corkage charge? If so, what is it?

❖ When are you open? For both lunch and dinner?

❖ Are children welcome?

❖ Can you cater for disabled people in wheelchairs?

❖ Are you air-conditioned?

❖ Do you have car-parking facilities?

❖ Do you cater for functions?

❖ How do I find your establishment?

Most reservations are taken over the telephone. A friendly and helpful telephone manner is essential, however busy you are, because potential customers are easily put off at this early stage and they may never call again.

❖ Always answer the phone promptly when it rings.

❖ Have the reservations book close to hand.

❖ State clearly the name of the establishment.

❖ Offer the caller your assistance (e.g. 'Good morning, may I help you?).

❖ Answer any questions clearly and politely. If you don't know the answer, find someone who does or offer to call back.

Before beginning to take the booking, make sure you have access to reservation sheets. The first things to be clarified are *when* the table is required and *how many* people there are in the party. Only after you have established that a suitable table is available when it is wanted, should you continue taking the reservation, asking for the following additional details:

❖ The host's name. (Ask for it to be spelt out if you are not sure.)

❖ The time of arrival.

❖ A contact telephone number.

❖ Any special requirements.

Confirm all the details by repeating the name, date, time of arrival, the number of people in the party and the contact phone number. Make sure all these details have been clearly written in the reservations book.

Complete the conversation with a show of hospitality—for example, 'Thank you, Mr Bryans. We look forward to seeing you on Thursday evening.'

Floor plans

The floor plans for a restaurant or a function are dictated by the number of covers (see below) and the style of the service to be offered. When these have been established, a floor plan is prepared by the dining-room supervisor as a guide for the set-up of the dining area and to assist in the seating of guests.

A floor plan is prepared using a simple outline of the floor space of the dining area that indicates the entrance and any other doorway or feature that could affect the placement of guests' tables and chairs.

Key points to consider when preparing a floor plan are:

❖ Position the tables so as to allow for sufficient movement by guests and service staff.

❖ Consider the placement of covers to avoid guests complaining of inappropriate placement—for example, near doorways, kitchens or toilets, behind pillars or in draughty areas.

❖ Accommodate guests' specific needs—for example, a business meeting, handicapped guests, honeymoon couples, family groups.

The plan is an effective guide in establishing the best use of the space and in meeting guests' specific requests for preferred tables.

Setting the mood/ambience

The total dining experience for the guests is much affected by the atmosphere created by management. The mood, or ambience, of the dining environment should reflect the time of day and the location, and create an atmosphere that is consistent with the desired character of the establishment.

Key points to be considered by management or staff in setting the dining atmosphere are:

❖ *Lighting.* Daylight or bright lighting is preferred for daytime meal services. Subdued light is more appropriate for evening dining. Candlelight can enhance the mood for evening dining but should not be used for daytime events.

❖ *Views.* Tables should be set to take best advantage of the views from the dining room (subject to the limitations of space).

❖ *Music.* Background music may be appropriate in establishing a mood. (In dining rooms where music is played, special consideration must be given to the placement of tables.)

❖ *Décor.* The decor should be consistent and create a harmonious atmosphere. Colour selection plays an important part in the dining experience. Some colours are warm, others cold; some are romantic, others businesslike, and so on.

While individual waiters may have no control over the colour and general decor of the dining room, they are often responsible for the details. Live plants and fresh flowers, for example, make a major contribution to the overall presentation and to the mood a room encourages. They must be carefully placed, well presented and well maintained.

How to cloth a table

Many different sizes and styles of table and tablecloths are used in the industry, and different ways of folding tablecloths are adopted by different laundries. There can therefore be no one correct technique of clothing tables. The tablecloth fold used in the following procedure for clothing a table is called a *concertina fold.* It is one of the more commonly used folds. The procedure as described assumes that the table has four legs, placed at the corners of the table.

CLOTHING PROCEDURE

❖ Check the table for steadiness and position it for ease of access for service. If the table is unstable, it must be stabilised.

❖ Stand centrally between two legs of the table.

❖ Position the folded cloth on the table with the two woven edges towards you and the two folds of the concertina facing away from you.

❖ Position the vertical centre crease in the centre of the table, holding the concertina fold.

❖ Lean across the table and release the bottom layer of the cloth to hang over the far edge of the table.

❖ Re-position the horizontal crease of the cloth in the centre of the table.

❖ Release the hold on the centre fold and draw the top fold towards you.

❖ Having centred the cloth both vertically and horizontally, the cloth should now be positioned with an equal drop all round, with the folds of the cloth covering the legs.

CHANGING A CLOTH DURING SERVICE

Cloths often have to be changed during food service, when guests are present at other tables and new covers are to be laid, or when there has been a serious spillage. In these circumstances, cloths must be changed with a minimum of fuss and, most importantly, without at any time exposing the bare tops of the tables to view. The procedure is as follows:

❖ Remove any articles remaining on the table to the sideboard.

❖ Stand centrally between two legs of the table.

❖ Holding the concertina fold, position the vertical centre crease of the clean cloth in the centre of the table.

Clothing a table before service

❖ Lean across the table and allow the bottom layer of the clean cloth to hang over the far edge of the soiled cloth.

❖ Take the soiled cloth in both hands, holding it between the little and ring (fourth) fingers.

❖ Concertina the soiled cloth towards you while opening the clean cloth above it.

❖ Fold the soiled cloth and remove it discreetly.

If you find that the table top is too wide for you to lean across and take the soiled cloth in both hands, as suggested above, try sliding the cloth towards you so that the far edge of the cloth only just covers the top of the table. While doing so, fold the cloth on your side of the table back towards the table so that no crumbs fall to the floor.

Station mise-en-place

Mise-en-place (French for 'put in place') can be defined as the equipment and food that is prepared ready for service before service begins. Station mise-en-place is the preparation of a waiter's workstation in a food-service area, housing all the equipment required for service.

A waiter's **station**, whether it is simply a clothed table or a special **sideboard** equipped with shelves, drawers and, sometimes, a hot box (plate/food warmer), should carry the following:

❖ all the necessary cutlery: for example, side knives, soup spoons, main (table) knives and forks, sweet spoons and forks, tea and coffee spoons

❖ service gear (tablespoons and forks)

❖ crumbing-down equipment

❖ service plates

❖ tea/coffee service equipment (milk jugs, sugar bowls, cups and saucers, teaspoons, etc.)

❖ glassware (tumblers, wine glasses—white and red)

❖ **underliners** (an underplate lined with a doily or napkin)

❖ bread service equipment (and butter—see below)

❖ napkins

❖ service trays

❖ toothpicks

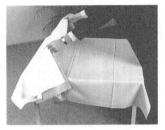

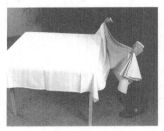

Changing a cloth during service

❖ menus

❖ wine lists

❖ spare docket books and pens (a docket and pen will be carried by the waiter)

❖ waiter's friend (usually carried by the waiter—see Chapter 13)

❖ condiments (sauces, pepper mill, etc.)

❖ clean table linen

❖ additional items to meet the specialist requirements of the establishment.

If the sideboard is equipped with a plate/food warmer, it must be turned on approximately 15 minutes before service begins.

Items from the service station used during service should be replaced or replenished during or at the end of the waiter's shift, or at the beginning of the next shift, as decided by the supervisor.

Preparing butter

Butter is prepared before the guests arrive. Chill individual portions of butter, whether they are curled, sliced or moulded, by placing them separately in iced water. This prevents the pieces of butter from sticking together. After they have been chilled, place the portions of butter on butter plates and keep them in the refrigerator until they are required for service. Take the butter plate to the table when you are serving the bread (see Chapter 5).

Preparing oil and vinegar

As an alternative to serving butter with the bread, some establishments prefer to serve oil and vinegar.

A portion of a good-quality olive oil at room temperature is placed in a small raised-edge plate and a drop of balsamic vinegar is added to the oil. In some establishments the oil and vinegar are served separately.

How to lay a cover

A **cover** may be defined as:

❖ A place setting at a table for one guest, laid to suit the type of menu offered.

❖ The number of guests to attend a function—'There will be 75 covers at the Rotary dinner', or to indicate the seating capacity of a dining area—'The Lawson Room seats 50 covers'.

There are two principal types of cover—à la carte/basic cover and set menu cover. The difference is explained below. Whatever the type of cover or shape of table to be laid, the following rules apply:

❖ All cutlery and glassware should be cleaned and polished before they are placed on the table.

❖ The main knife and fork should be positioned 1cm from the edge of the table and 25–29cm apart (depending on the size of the establishment's dinner plates).

❖ Side plates are always positioned to the guest's left.

❖ Side knives are placed on the side plate, to its right-hand side and parallel with the main knife and fork, so that a bread roll can be placed on the plate.

❖ The blades of all knives on the cover should face left.

❖ The first or only wine glass is positioned 2.5cm from the tip of the main knife. Additional glassware is positioned at a 45° angle to the left of the first glass (see below).

❖ A folded napkin is placed in the centre of the cover (see below).

Types of cover

À LA CARTE/BASIC COVER

An à la carte menu features a variety of dishes individually priced. The guests select the dishes they would like, usually up to and including the main course.

The basic à la carte cover—the lay-up performed before the guests arrive—is for a main course only. After the guests have ordered, this basic cover is then corrected to suit the customers' actual orders (see Chapter 6).

À la carte cover/basic cover

An à la carte cover includes:

❖ main knife and fork

❖ side plate

❖ side knife

❖ table centre items (including bud vase or candlestick, cruets, tent card and table number)

❖ wine glass

❖ napkin.

Set menu cover, with no dessert gear

SET MENU COVER

A set menu features prearranged items at a fixed price for the whole meal. Because it is known in advance what will be served to the guests, the cutlery and glasses for the whole meal are laid in advance. The cover illustrated is for a set menu offering a plated appetiser, soup, a fish entrée and a main course.

Set menu cover, with dessert gear

The cutlery required for this menu is:

❖ entrée knife and fork

❖ soup spoon

❖ fish knife and fork

❖ main knife and fork

❖ side plate

❖ side knife

❖ glassware

❖ napkin.

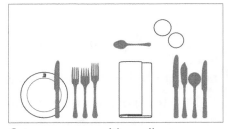

Set menu cover, with small spoon

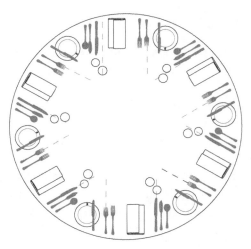

Placement of covers in a round table set-up (set menu)

Note that the various items of cutlery are set so that the cutlery used for the first course is outermost, and the remaining cutlery is set in the order in which it will be used, working inwards from the outside.

BISTRO COVER

A bistro cover is a simplified setting to suit a less formal style of dining. The simplicity of this cover makes it suitable for buffets, counter meals and barbecues.

The cutlery used and the exact placement of the equipment will vary according to the menu and the style of the establishment.

Dessert cutlery

Dessert gear (the dessert spoon and fork) is not usually laid before the meal begins, but corrected (placed) after the main course has been cleared. It may also be set across the top of the cover before the meal. This alternative will require the waiter to move the fork and spoon down just before the course is served (see Chapter 6).

As an alternative to laying the dessert spoon and fork prior to the meal, we suggest that dessert gear should be taken to the table on a service plate and laid only when needed for the sweet course. This will allow additional room on the table and ensure that the cutlery will be absolutely clean when the time comes for it to be used.

Note that not all desserts require a spoon and fork; some may require a fruit knife and fork, or a small spoon.

Glassware

The wine glasses, like the cutlery, are set in the order in which they will be used. They are usually set in at an angle of 45° from the first glass to be used, which (as already noted) is placed about 2.5cm from the tip of the main knife. In the diagram the glasses are set at 45° inwards to the left (towards the centre of the cover) so as not to confuse the guests as to which glassware is theirs; in special circumstances the glasses may also be set at 45° outwards to the right of the cover.

If a water glass is to be set, the glass should be positioned to the right of the first wine glass.

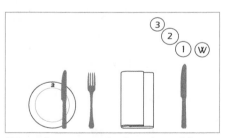

Wine glasses: standard setting with water glass

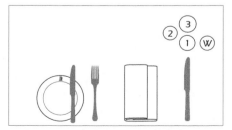

Wine glasses: triangle setting with water glass

Should the table not have sufficient room for the glassware to be set in a row, it may be set in a triangle.

Napkin folds

A folded table napkin is placed on the table for the guest's use and to contribute to the appearance of the cover and the whole dining environment. The way in which the napkin is presented depends on the type of establishment and the type of service.

It is an advantage if napkin folds are kept simple, as less handling is involved. Less handling makes for more hygienic napkins (as well as being less time consuming). However, some establishments require more elaborate folds for aesthetic reasons.

Detailed on the following pages are examples of some commonly used professional napkin folds. Either starched linen or paper napkins can be used for folding.

A professionally folded napkin will stand by itself without the aid of cutlery or glassware.

Sail   Bistro   Cone   Bishop's hat/mitre

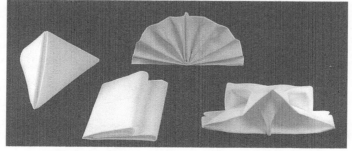

Inverted sail Envelope   Fan   Five star

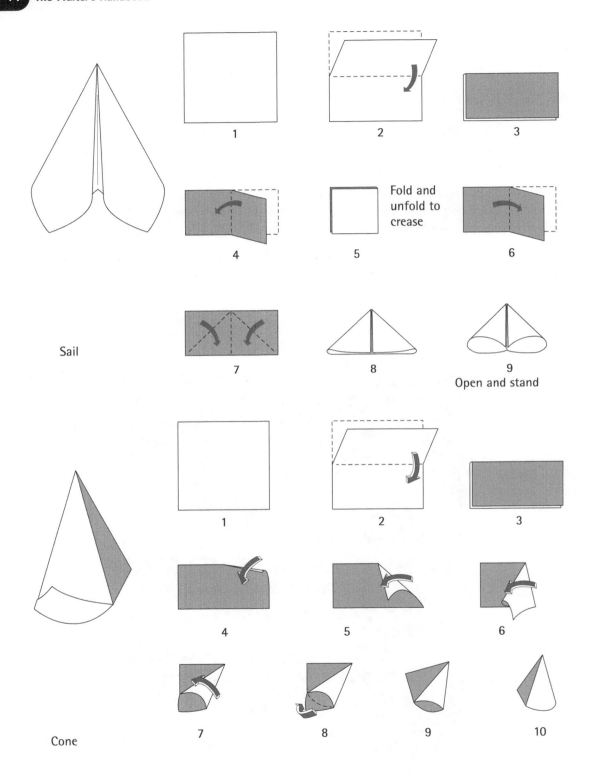

1

2

3

4

5

Fold and
unfold to
crease

6

Sail

7

8

9

Open and stand

1

2

3

4

5

6

7

8

9

10

Cone

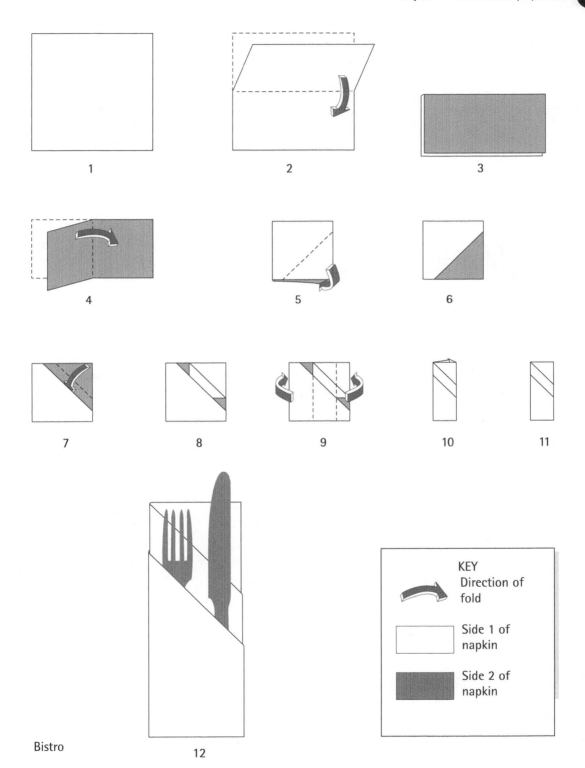

1

2

3

4

5

6

7

8

9

10

11

Bistro

12

KEY
Direction of
fold

Side 1 of
napkin

Side 2 of
napkin

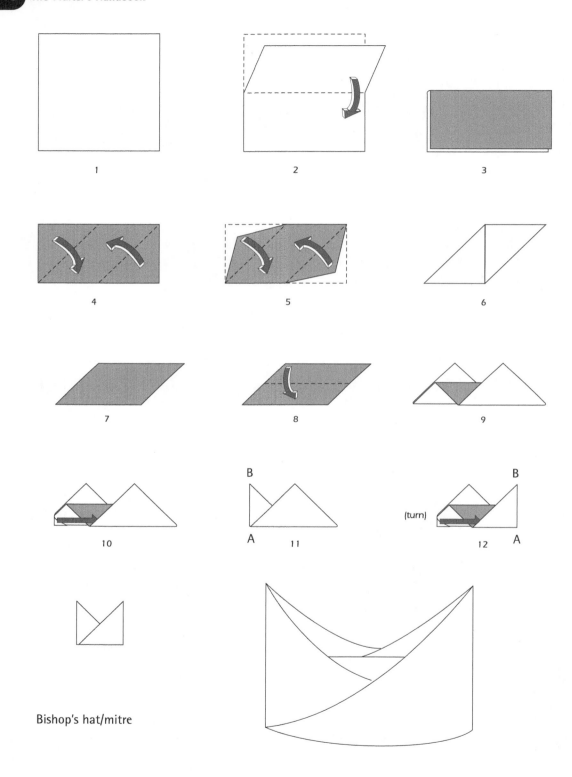

Bishop's hat/mitre

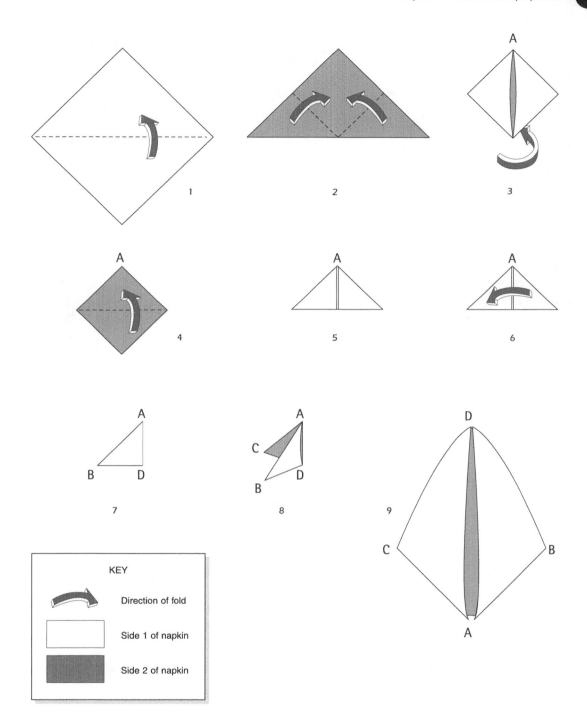

KEY

Direction of fold

Side 1 of napkin

Side 2 of napkin

Inverted sail

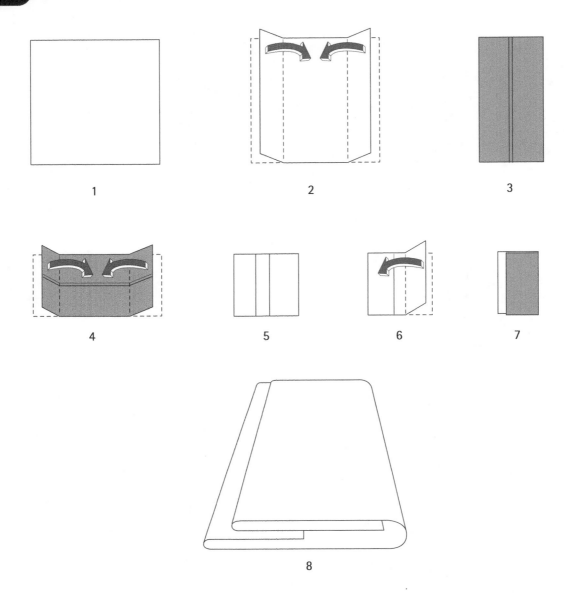

1

2

3

4

5

6

7

8

Envelope

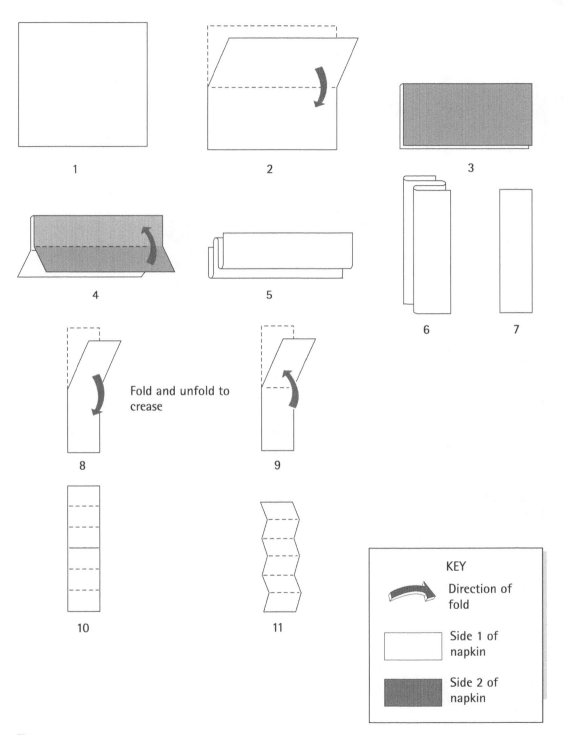

Fold and unfold to crease

KEY

Direction of fold

Side 1 of napkin

Side 2 of napkin

Five star

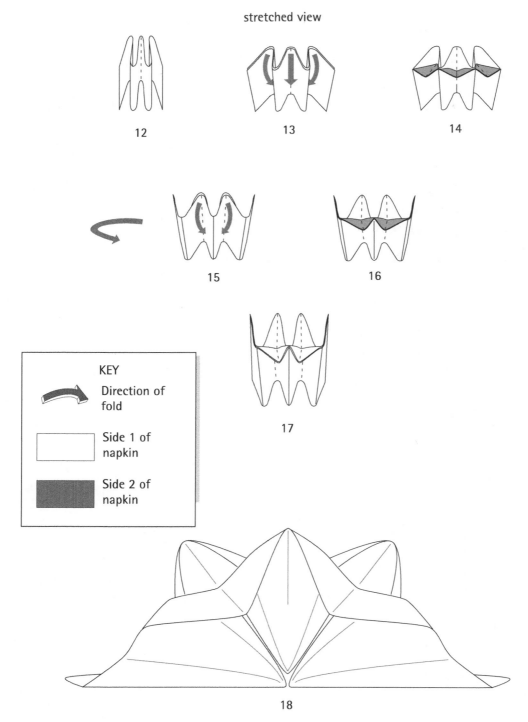

stretched view

12

13

14

15

16

17

KEY

Direction of fold

Side 1 of napkin

Side 2 of napkin

18

Five star (continued)

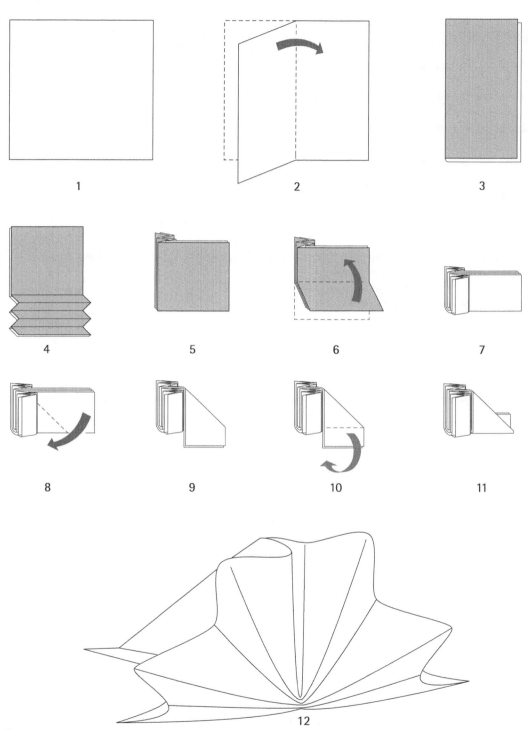

Fan

QUESTIONS

1. Why are high standards of presentation essential?

2. What are the two questions you need to ask before you take a reservation?

3. What additional details are needed to complete the reservation?

4. What are three important points to take into consideration when preparing a floor plan?

5. What are four key points in creating the right atmosphere?

6. How can you change a soiled tablecloth without exposing the table top to the guests?

7. Waiters should always check their stations before service starts. What items should they check for?

8. How is an à la carte cover set?

9. How is a table d'hôte cover set?

10. What are two alternative ways of positioning the wine glasses?

11. Demonstrate folding all of the napkins (pages 44–51) without looking at the diagrams.

Chapter 5

Food service procedures: the preliminaries

The objective of food service is to meet the needs of guests and to ensure that their dining experiences are both pleasurable and memorable. Why then should a guest be inconvenienced by the service? A guest who has to duck and dive throughout a meal to make way for the service will only remember that the service was not pleasurable.

On completion of this chapter you will have a good understanding of the following:

❖ The order of service

❖ Greeting and seating guests

❖ Opening napkins

❖ Service of water

❖ Service of bread

❖ Taking orders for pre-dinner drinks (apéritifs)

❖ Serving pre-dinner drinks.

Food service procedures vary considerably throughout the industry. Whether an establishment offers a traditional style of service, modern service or its own particular variation is of little importance; what matters is that the establishment is consistent in the services it offers.

Because of the great variety of food service procedures in use today, many people entering the industry for the first time become confused about what are the 'correct' practices. The answer is that no one set of techniques is correct in all circumstances. What we have tried to do in this and the following chapters is to present commonsense solutions to the various problems and challenges of service that will be applicable in most circumstances. However, the reader must always be aware that in some establishments other techniques and practices may be required.

A good example is the much-debated question of whether service should be clockwise around the table or anti-clockwise. It doesn't really matter whether an establishment chooses to serve in a clockwise or anti-clockwise direction. What does matter is that, whatever the direction adopted by an establishment as its standard direction of service, that standard should be followed by everyone working in the establishment. Consistency is what matters.

The order of service

The 'order of service' is a sequential checklist of services from the arrival to the departure of the guest. It will differ in detail depending on the style of the establishment and the services it offers.

The sequence of service will take into account the particular tasks to be performed to achieve a smooth flow of service to suit the special needs of each guest.

Many points must be considered when an order of service is defined: for example, whether iced water will automatically be offered, whether hot or cold towels will be provided, and how the food and beverage services are to be coordinated.

If several groups of guests arrive simultaneously at tables that are the responsibility of a single waiter, then it is necessary to read the differing needs of the various groups and to adjust the order of service to ensure that the tasks can be performed without overloading the station.

Checklist of service

A written checklist of service ensures the consistency of the services offered and acts as a guideline to part-time and new waiting staff. *What follows is an example (and only an example) of such an order of service* from the time the guests are taken to their table.

❖ Greet and seat the guests.

❖ Open the napkins.

❖ Offer iced water.

❖ Take pre-dinner drink orders.

❖ Serve pre-dinner drinks.

❖ Serve the bread and butter.

❖ Offer the menu, suggest specials, and inform the guests of variations to the menu.

❖ Take the food order up to and including the main course.

❖ Offer the wine list.

❖ Record and transfer the food order to the kitchen.

❖ Take the wine order.

❖ Serve the wine.

❖ Correct the covers, up to and including the main course.

❖ Serve the first course.

❖ Clear the first course.

❖ Top up wines and open fresh bottles as ordered.

❖ Serve additional starter courses (for example, a second entrée).

❖ Clear the course preceding the main course.

❖ Call away the main course (that is, request of the kitchen that it be prepared and finished for service).

❖ Serve the salad.

❖ Serve the main course.

❖ Enquire (after the guests have had the opportunity to taste the food) whether the meals are satisfactory.

❖ Clear the main course.

❖ Clear the side plates, salad plates and butter dishes.

❖ Brush/crumb down.

❖ Offer hot (or cold) towels.

❖ Offer the wine list for the selection of dessert wines (or, if the guests prefer it, continue to serve the wine selected earlier).

❖ Offer the menu for dessert, suggesting specials, and inform the guests of variations to the menu.

❖ Take dessert or cheese order.

❖ Record and transfer the dessert order to the kitchen.

❖ Correct the cover.

❖ Serve the dessert wines or other beverages selected.

❖ Serve the dessert or cheese course.

❖ Take the order for coffee/tea. (The coffee/tea may be served with the dessert/cheese course if requested by the guests or as a separate service.)

❖ Record the coffee/tea order to the cashier docket.

❖ Take the after-dinner drinks order.

❖ Correct the cover.

❖ Serve the after-dinner drinks.

❖ Serve the coffee/tea.

❖ Serve the petits fours.

❖ Prepare the guests' account.

❖ Offer additional coffee/tea.

❖ Present the guests' account when it is requested.

❖ Accept payment and tender change.

❖ Offer additional coffee/tea.

❖ Farewell your guests.

Greeting and seating guests

First impressions are extremely important. Guests arriving at a restaurant gain their first impression of the establishment substantially from the willingness of the staff to acknowledge their presence and the greeting they receive. If

Offer the guest a chair

the greeting is both warm and efficient, guests will immediately feel that they can expect the rest of their experience to be pleasurable, and they will feel confident that they will be in the hands of reliable professionals.

The waiting service begins with the greeting and seating of the guests.

In larger establishments, guests may be received by a head waiter or supervisor and taken to their table (after checking reservations, etc.). There they are introduced to their table waiter, who takes over responsibility for their service. In smaller restaurants, a single waiter will be responsible for the whole operation. In both cases, the procedure is as follows:

❖ Acknowledge new guests as soon as they arrive.

❖ Approach the guests with an appropriate welcome—for example, 'Good evening.'

❖ If the guests wish to eat, ask whether they have a reservation. Check the reservation. If no table has been booked, check that one is available.

❖ When checking the reservation, note the host's name—the table will usually have been reserved in the name of the host. It is important to establish who the host is. (The host may, of course, be female or male.)

❖ Show the guests to their table.

❖ Offer the guests a chair to encourage them to be seated.

Opening napkins

Opening the napkins for your guests ensures that the napkin is out of the way when drinks and food are served. Some guests will open their own napkin as soon as they sit down; others will wait for you to open it for them. The technique is:

Opening a napkin

❖ Pick up the napkin with the right hand from the guest's right.

❖ Shake the napkin from its fold into a triangle.

❖ Place it across the guest's lap with the longest side of the triangle closest to the guest.

❖ Move around the table opening the napkins; open the host's last.

Water service

Iced water may be offered to the guests after the greeting/seating procedures. The purpose of serving iced water is to refresh the guests' palates and allow them time to select a pre-dinner drink.

Iced water is a valuable addition to the meal experience and is appreciated by guests. It should always be available, although in some establishments it may not be the practice to serve it unless it is asked for. (Some countries, including the United States and Japan, require fresh water to be made available, and visitors from those countries will expect iced water to be available without their having to ask for it.)

The procedure for serving water is:

❖ The water glass is positioned to the right of the wine glass above the table knife.

❖ Water is poured from the guest's right.

❖ Move around the table pouring the water, serving the host last.

❖ Continue to offer water throughout the meal, as required.

In some establishments jugs of iced water may be left on the table for the guests to help themselves.

Serving iced water

Bread service

Bread, in some form or other, is usually served as soon as the guests are seated. It may be placed in a basket on the table or served individually (silver service). The advantage of the individual service of bread is that it leaves more room on the table. The silver service technique is:

❖ Carry the bread basket, and butter or oil and vinegar plate, in the left hand, using the two-plate method (see Chapter 7).

❖ Place the butter or oil and vinegar plate in the centre of the table.

❖ Transfer the bread basket to the flat of the left hand.

❖ Serve from the guest's left.

❖ Hold the left hand (with the bread basket) down over the edge of the side plate.

❖ Transfer the bread from the basket to the side plate, using service gear (see Chapter 8).

❖ Move around the table, serving the host last.

Pre–dinner drinks/apéritifs

Pre-dinner drink (apéritif) orders should be taken as soon as possible after guests have been seated. The waiter should encourage the guests to try something a little adventurous or different by suggesting speciality cocktails or beverages, allowing them time to consider their preferences.

❖ Offer the drinks/cocktail list, or suggest a variety of the beverages available.

❖ Assist the guests in making their selections by explaining what is in the various cocktails and what they look like.

❖ Record the orders in sequence around the table.

❖ Note any special requirements (no ice, etc.).

❖ Record the sales following the house control system.

❖ Place the order with the bar.

SERVING PRE-DINNER DRINKS

❖ Arrange the drinks in sequence of service on a drinks tray.

❖ Carry the drinks to the table on the drinks tray.

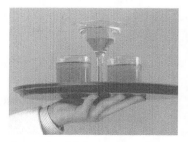

Carry the drinks to the table on the tray

❖ Hold the tray on the left hand away from the table (see Chapter 15).

❖ Serve the drinks in sequence around the table, serving the host last.

❖ Place each drink to the right of the guest's wine glass.

Serving pre-dinner drinks

Your guests should now be comfortably settled and ready to turn their attention to the menu.

QUESTIONS

1. What is the purpose of the 'order of service'?

2. Why is it important to establish who is the host?

3. Describe the procedure of opening napkins.

4. Why serve iced water?

5. When and how is bread served?

6. How can guests be encouraged to order pre-dinner drinks?

7. In what order should apéritifs be served?

ACTIVITY

Create a checklist of service that identifies the points of service that relate to a specific workplace operation.

1. Describe the type of operation.

2. List the specific points of service, in sequence, that relate to this operation. The number of points listed will relate to the type of operation and the services offered.

Chapter 6

Taking the orders and correcting the covers

Guest: What would you recommend today?
Waiter: Everything is good!

Well, that was a big help, wasn't it?

When you have completed this chapter you will have a thorough understanding of the following:

❖ Presenting the menu

❖ Food sales and sales techniques

❖ Taking food orders

❖ Dietary and cultural awareness

❖ Use of a service plate

❖ Correcting the covers.

Once the guests are comfortably settled and have been given their pre-dinner drinks and bread, they are ready to turn their attention to the menu and select what they want to eat. The menu is offered, the order is taken and the cover is corrected.

Presenting the menu

A waiter should not simply take orders and serve what is ordered. The waiter's job is more proactive—waiters should make things happen. They are salespeople as well as service people.

Presenting the menu is a time for suggestive selling. The waiter has the opportunity to actively sell items on the menu, and 'specials' and side dishes that may not be on it. At no other time does the waiter have so much of the guests' attention, and it is an opportunity not to be missed.

Before presenting the menu you must understand all the items on it and be able to describe how they are cooked and served. (You may find the glossary helpful for this.) You must also know the details of the daily specials.

Menus should be offered in such a way as to encourage the guests to select their meals reasonably quickly, without appearing to put any pressure on them to do so.

Presenting the menu

DIFFERENT FORMS OF MENUS

Menus come in different forms. In traditional restaurants the menu is usually presented in a cover, but in less formal establishments it can be written on a blackboard, or a card, or a souvenir place mat—there is no limit to the possibilities.

Particular venues may well have devised their own methods to suit the form in which their menus appear and the style of the operation. These special house rules must be followed by the waiting staff.

THE TECHNIQUE OF PRESENTING A MENU

If the menu is in a cover, it should be opened before it is presented to encourage the guests to read it and make their selections.

❖ Carry the menu on the flat of the left arm.

❖ Open the menu from the top with the right hand.

❖ Present the menu to the guest's right.

❖ When all the guests have received a copy of the menu, suggest items that don't appear in the menu and mention any variations to the menu items.

Describing and recommending dishes

You are now likely to be asked questions about the specials, and about the items on the menu. You must be able to describe the dishes and how they are cooked and served, concisely, accurately and attractively.

You may also be asked to make recommendations. Be prepared to assist the guests in making their selections. To say 'Everything is good!' isn't helpful. Establish what sort of dish the guest wants—fish or meat, hot or cold—and then direct him or her to those dishes that seem most appropriate.

THE WAITER AS SALESPERSON

This is the time when the waiter's skill as a salesperson comes into play. 'Hard sell' techniques are seldom effective. Sales are made by suggesting items that the guests might well have ordered had they known of them. What you are providing is better service—making the guests' experience more complete and enjoyable—rather than a series of sales pitches. You might helpfully say, for example, that the fish of the day is fresh from the market, or that buffalo steaks are a new menu item and have proved very popular.

It is a basic sales technique not to invite negative answers. Instead of saying 'Would you like a starter?', which invites the answer 'No', ask instead: 'What would you like to start with?'

Some establishments (in particular, American-based chain restaurants) give detailed instruction to their waiting staff in sales techniques; in others, the whole emphasis is on helpful service without any mention of the word 'sales'—'sales' might almost be thought of as a dirty word—but the effect of good service should be the same in either case: contented guests and profitable food sales for the establishment.

To be an extremely effective salesperson, the waiter need only be sincerely helpful, friendly, attentive and enthusiastic, with a thorough knowledge of the menu and the ability to describe and recommend suitable items.

Special requests, or dietary or cultural requirements

The ever-increasing demand for special foods, food preparation and service, be it for cultural or dietary reasons, has increased the need for the waiter to become well versed in the terminology, its meaning, and the implications for guests if their requests are not fully appreciated and met.

Today's customer demands a high standard of service. When you are able to answer customers' questions, you will go a long way to satisfying their needs and making your job more challenging and rewarding.

DIETARY AWARENESS

Many guests today have special dietary requirements that include food allergies and intolerances, and therapeutic or lifestyle diets.

❖ A **food allergy** is an immunological reaction to food proteins.

❖ A **food intolerance** is a pharmacological reaction (like side effects from a drug) to the chemicals in foods.

❖ A **therapeutic diet** can meet the nutritional needs of the sufferer of a medical illness or condition.

❖ A **lifestyle diet** is a preference to eat or reject certain foods for various reasons.

Let's take a look at each of these.

Food allergies

Food allergies can be very serious and can lead to hospitalisation and in some cases death. Some people with peanut allergies cannot even walk down the chocolate bar aisle of a supermarket without having a reaction. Never be too complacent when a guest says they have an allergy. Check and double check with the kitchen regarding the ingredients and possible contamination of allergy foods in the dish requested by the guest.

Some of the more common food allergies are:

❖ *Wheat.* Some guests cannot eat any food that contains wheat. This includes bread, pasta, cereals or any food containing gluten or food starch. Most foods have some traces of wheat, and severe allergy sufferers must buy produce from stores that sell wheat-free products. Eating out can be difficult.

❖ *Milk.* Guests may be allergic to all milk—cow, sheep, goat and sometimes soy—as they all contain a milk protein. Many dishes contain milk or products such as butter that are made from milk. Milk allergies exclude from the diet foods such as pastries, cakes, some breads, biscuits, processed cereals, pancakes, cream soups, custards and chocolate.

❖ *Peanuts*. Avoiding nuts can be difficult because many foods may contain traces of nuts as they are manufactured on the same production line as foods containing nuts—for example, chocolate bars. Severe allergies to nuts can lead to a severe reaction. Many people with a severe allergy will carry self-medication, just in case. Indonesian, African, Chinese, Thai and Vietnamese dishes often contain peanuts, or are contaminated with peanuts during preparation. Additionally, foods sold in bakeries and ice cream shops are often in contact with peanuts.

❖ *Fish/shellfish*. Fish-allergic guests should be cautious when eating away from home. They should avoid fish and seafood restaurants because of the risk of contamination in the food-preparation area of their 'non-fish' meal from a counter, spatula, cooking oil, fryer, or grill exposed to fish. Also, fish protein can become airborne during cooking and cause an allergic reaction. Some individuals have had reactions from walking through a fish market.

❖ Other food allergies may include soybean, tree nuts (walnut, cashew, pistachio) and eggs.

If a guest lets you know they have an allergy, check with the chef to find out what ingredients are contained in the dish. If in doubt, suggest an alternative to the guest.

Food intolerance

'Food intolerance' is the term used to describe a reaction to a food component. A pharmacological reaction refers to the drug-like side effects caused by chemicals in food as natural or added components. For example, asthmatics are most likely to be affected by sulphite preservatives in a wide range of foods and drugs, including fruit-flavoured cordials and drinks, wine, bread, sausages and dried fruit, and some medications. Many people are also sensitive to monosodium glutamate (MSG), a natural flavour enhancer that is found in many takeaway foods, parmesan cheese, anchovies, miso, and sauces such as soy.

Unlike a food allergy, where a person may need to completely avoid the food(s), most people can tolerate some amount of the food or chemicals of which they are intolerant. They can prevent symptoms by avoiding an accumulation of the food chemicals in the body over several days.

Other dietary needs

While a guest may not have an allergy or intolerance, they may opt not to eat some foods. In restaurants today, guests are requesting the following types of meals:

❖ *Fat free.* To be classified as fat free, food items must be less than 0.15 per cent fat. Different types of fats react differently inside the body. Saturated fats (found mostly in animal products) increase blood cholesterol, which is a risk factor in coronary heart disease. Mono-unsaturated and poly-unsaturated fats tend to lower blood cholesterol. Butter, coconut and palm oil, cottonseed oil, lard, cocoa butter and beef tallow are all high in saturated fats. These fats are commonly found in many fast foods and in commercial products such as biscuits and pastries.

❖ *Carbohydrate free.* Food provides fuel for our body in the form of fat, protein, carbohydrates and alcohol, but carbohydrates are the body's preferred fuel source. Carbohydrate-containing foods include bread, breakfast cereals, rice, pasta, legumes, corn, potato, fruit, milk, yoghurt, sugar, biscuits, cakes and lollies.

❖ *Vegetarian or vegan* (a diet free from any animal products, including eggs and dairy products). Vegetarian food is not so much a request these days as a necessity on any menu. Where once a menu may have contained one or two vegetarian options, restaurants these days choose to serve vegetarian-only meals as the customer demand increases.

❖ *Low in cholesterol.* There is no need to eat foods that contain cholesterol; your body can produce all the cholesterol it needs. High-cholesterol foods are usually foods high in saturated fats. They include full-fat dairy products, processed meats such as salami, snack foods such as chips, pastries and cakes, and deep-fried takeaway foods.

❖ *Organic.* Organic food is grown and produced without synthetic chemicals (such as pesticides or artificial fertilisers). It does not contain genetically modified (GM) components, and is not irradiated. Organic produce can include fruit and vegetables, meat and meat products, dairy foods, dried legumes, grains, eggs, honey and some processed foods.

For further information, see the Better Health website: www.betterhealth.vic.gov.au

Therapeutic diet

A therapeutic diet is a selection of foods and cooking methods that is used to act as a preventative, supportive or controlling measure to meet the nutritional needs of the sufferer of a medical illness or condition. There are a great number of medical problems that require a modified diet. Diabetes and heart disease are just two examples.

Some illnesses lead to disturbances in metabolism. Some foods may be harmful in some people's bodies, causing allergic reactions such as cramps, internal swelling or loss of consciousness. However, some illnesses may be controlled or even cured by altering the metabolic process. In these cases, the regulation of the food consumed is a very important part of the cure.

Medical conditions that require special diets include digestive organ diseases and coeliac disease.

Lifestyle diets

Lifestyle diets are food requirements for people who choose either to eat or to reject certain foods for reasons of health, morals, finances or personal taste—for example, vegetarian and weight reduction diets.

Vegetarians are people who, for various reasons, don't eat the flesh from animals. There are several types of vegetarian diets.

❖ *Lacto-ovo*. This dietary style gets its name from the Latin word '*lacto*', meaning 'milk', and the Italian word '*ovo*', meaning 'egg'. Foods eaten include eggs, cheese, milk, cream, yoghurt and butter. Foods not included in the diet are meat, fish and poultry.

❖ *Lacto*. This vegetarian diet is similar to the lacto-ovo diet; however, eggs are not eaten. Foods eaten include cheese, milk, cream, yoghurt and butter. Foods not included in the diet are meat, fish, poultry and eggs.

❖ *Vegan*. Vegans are strict vegetarians. Foods eaten include plant products, nuts, soy products, tofu and molasses. Foods not included in the diet are meat, fish, poultry, eggs, milk, cheese, butter, gelatine and lard. Strict vegans also reject items such as honey, because it is collected from bees, and margarine because it contains milk powder. Catering for a vegan requires additional thought to ensure that no animal products are incorporated in the meal or used in the cooking process.

❖ *Piscatorian*. The Piscatorian style gets its name from the star sign Pisces (fish). Foods eaten include fish (all seafood), cheese, milk, cream, yoghurt, butter and eggs. Foods not included in the diet are meat and poultry. In the strict sense of the word, Piscatorians are not true vegetarians.

It is very good customer service if you know your menu thoroughly and can guide guests to items on the menu that are low in fat, carbohydrate free, or vegan. Similarly, letting the guest know that you use free-range eggs or chicken on your menu may allow them to make a menu choice they may not have ordinarily made.

CULTURAL APPRECIATION

Different religious beliefs mean that customers will only eat certain foods or food prepared in certain ways. Most customers will let you know if they have a special need when making a booking or arriving at the restaurant. These needs may vary according to how strictly they follow their religion.

Cultural foods may include:

❖ *Kosher.* Food that complies with Jewish dietary law is referred to as 'kosher'. Certain animals are considered 'unclean' and are not eaten—for example, fish with no scales or fins, all molluscs and crustaceans, pigs, and animals living underground. All edible animals must be properly slaughtered, soaked, salted and washed according to kosher standards. Meat and milk products cannot be prepared or eaten together. Many international hotels have now built kosher kitchens where food preparation is overseen by a Rabbi (spiritual leader of a Jewish congregation).

❖ *Halal.* Food that is slaughtered and prepared according to Islamic law. Muslims don't eat pork or decaying meat, or the meat of animals that have died violently. Muslims also fast for the month of Ramadan, when they don't eat or drink between sunrise and sunset. Ramadan occurs at varying times each year according to the lunar calendar.

❖ *Prashad.* Food that is blessed for Hindus. Hindus revere (hold in high regard) the cow, and many Hindus are vegetarians. While they may eat other meats, they don't generally eat beef.

Understanding different cultural and religious food requirements will help you to ensure that guests' needs are respected. If in doubt about what the guest can or cannot eat, always ask. Some people practise their religion to varying degrees and may have special requirements when eating in your restaurant or café.

FOOD FOR THOUGHT

There are many different types of guests and they all have varying needs. Some people's health depends on what they eat; others may be guided by lifestyle or religious considerations. An important part of your role is ensuring that these diverse needs are catered for.

The menu used in your establishment is a perfect starting point when preparing foods for specific therapeutic, cultural or religious needs. First look through the menu for dishes that may be suitable for the requirement. If there are no dishes on the menu that suit in their unaltered form, consider dishes that may be modified by adding or omitting ingredients or by simply using an alternative cooking method.

When catering to the needs of guests with special dietary requirements, try to put yourself in their shoes. They don't want to cause a fuss or create extra work for you. They only require a meal that is safe, healthy and appetising to eat. Try to do whatever you can to make their dining experience enjoyable and uncomplicated.

Children

Young children may get bored and unruly if not attended to, making things difficult for both the waiter and the parents. Give the children something to eat or do as quickly as possible: paper, pens, crayons if you can manage it, or at the very least some bread and butter to eat. They must be served as soon as possible. You may offer to serve their main course when the adults are served their entrées.

Taking and placing food orders

Food orders are taken as soon as the guests have made their selections. You must be alert to the signs that the guests are ready to order so that they are not kept waiting. They may, for example, close their menus and place them on the table.

ORDER-TAKING TECHNIQUES

To ensure prompt service and fulfilment of the guests' orders, the waiter must record all the necessary information so that there is no doubt which guest ordered what.

Orders can be taken in various forms, depending on how many guests there are at the table, and the procedures of the establishment. Some establishments have preprinted forms that simply have to be ticked. Often the waiter will have to use a blank docket.

The way the order is taken must serve three functions. It must be clear to the kitchen what dishes are to be prepared (and whether there are any special requirements). The waiter must be able to see which dish is to be served to which guest. And when the bill is prepared, it must be clear what has been ordered and consumed. *Clarity* is essential.

The great enemy of clarity is the use of personally devised abbreviations. If abbreviations are used, they must be consistent and in a style agreed by *both* the waiting and kitchen staff. Avoid letter abbreviations; use at least part of the name of the dishes ordered—according to the agreed style.

There are a great many different systems for order taking in use. Waiters new to an establishment will usually be trained in that establishment's ordering systems as a part of their induction and orientation. If you are not given formal training in the ordering system in use, you must in any case make sure that you thoroughly understand it. Systems may include the use of hand-held units, swipe cards, electronic billing, remote printers, and a variety of innovative technology in place of simple handwritten dockets. The docket systems described below are just two of the many possibilities.

PROCEDURE FOR ORDER TAKING

❖ Make sure that the guests are ready to order.

❖ Take the order of the guest seated next to the host first, and work around the table, finishing with the host's order.

❖ Take the order up to and including the main course.

❖ Note any special requirements (for example, a special dietary requirement, such as no milk to be used in the preparation of the dish, or a service requirement, such as meals required very quickly because the guests are going on to a show).

❖ Repeat the order to the guests to make sure that it is correct.

❖ Transfer the order to the kitchen docket (using either a manual or a computer system), including the special instructions.

❖ Record the sale for billing purposes, following the house control system.

❖ Place the order with the kitchen.

Docket systems

Control or docket systems are used to:

❖ provide the cashier with the information to make up the bills

❖ keep a record of all food and drink used

❖ keep a check on stores so that wastage can be minimised and pilferage discovered

❖ store information so that the cost of each menu item can be calculated accurately and the profit made on it worked out

❖ provide a breakdown of sales and financial statistics.

There are many different docket systems in use in catering establishments, but they can be divided into four main types:

❖ triplicate docket system—a traditional manual system often used in medium and large-sized hotels and restaurants

❖ duplicate docket system—often used in small, informal restaurants

❖ electronic billing machines—used in some establishments where the waiter, rather than the cashier, prepares the bill

❖ computerised systems—used in large establishments and chain restaurants.

The sample manual system illustrated on pages 73–74 is another option, with elements of both standard, duplicate and triplicate systems.

TRIPLICATE DOCKET SYSTEM

Three copies of the docket are made: a top copy for the kitchen, a second copy for the cashier and a third for the waiter.

The docket must be clearly written in the same language as the menu to avoid misunderstandings. Only agreed abbreviations should be used. The docket must contain:

❖ table number

❖ number of covers

❖ date

❖ waiter's signature.

Any cancellation of a docket must be authorised by the head waiter or supervisor.

Often more than one docket is needed for a meal: for instance, the dessert often requires a second docket to be written. In these cases, the docket should be headed 'Following' to show that it is the second docket for a particular table. Three copies of this docket are needed as with the original docket.

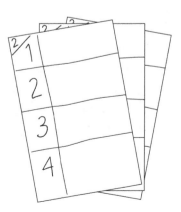

If an extra portion of food is required, because not enough portions were sent from the kitchen, a docket

should be issued headed 'Supplement'. This docket should be signed by the supervisor or head waiter. There is usually no charge for such orders.

If there is an accident with a dish, and a docket is written for a repeat order, it should be headed 'Accident'. This docket should also be signed by the head waiter or supervisor and no charge should be made. The same procedure is followed with the three copies.

If the wrong dish is sent from the kitchen, it should be returned with a docket headed 'Return'. The name of the correct dish and the returned dish should be written on the docket.

DUPLICATE DOCKET SYSTEM

This system uses only two copies of a docket. It is normally used in establishments offering a limited menu and may be preprinted. It may also make use of perforated strips, each one for a different course. The waiter tears off the strip and sends it to the kitchen as required. The docket should contain the following information:

❖ serial number of docket pad

❖ waiter's code number identification

❖ table number

❖ time the order is placed

❖ date.

The duplicate copy is also used as the bill.

The waiter must ensure that all items are entered on the bill. If the waiter presents the bill and the guests pay the cashier, an analysis of the waiter's takings will be drawn up by the cashier.

Sample of a simple manual order-taking docket system

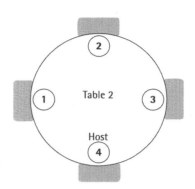

WAITER'S ORDER PAD

A small plain jotter pad may be ruled up by the waiter to accommodate the orders of all guests at each table.

Indicate table number →

Indicate position of guests at the table by numbering them down the page, starting with the guest seated next to the host, moving around the table. The host is the last person to order.
(See diagram of table.)

Note re host.
(Waiters sometimes indicate a characteristic of the host on their ordering pad to assist in remembering the order of service.)

2 /		
1	Soup	Steak Medium salad only
2	Paté	Veal
3	Soup	Fish
4 Bow tie	Prawns	Steak MR No sauce

ESTABLISHMENT CONTROL DOCKET

Duplicate docket with carbon. The orders are transferred by the waiter from the order pad to the control docket. One copy goes to the kitchen; one goes to the cashier.

This system allows waiters to make quick notes in a way most helpful to them, while preserving all the controls needed by the management.

Rule off the close of each course →

THE UNICORN 0241

2 soup	
1 paté	
2 prawns	
2 steak	1 MR – NO sauce 1 med – salad only
1 veal	
1 fish	

DATE	TIME OF ORDER	SERVER	NUMBER OF GUESTS	TABLE NUMBER

Control serial number

Using the waiter's order pad to ensure correct:
❖ ordering
❖ placement of gear
❖ placement of food.

The order for each guest is written across the pad up to and including the main course.

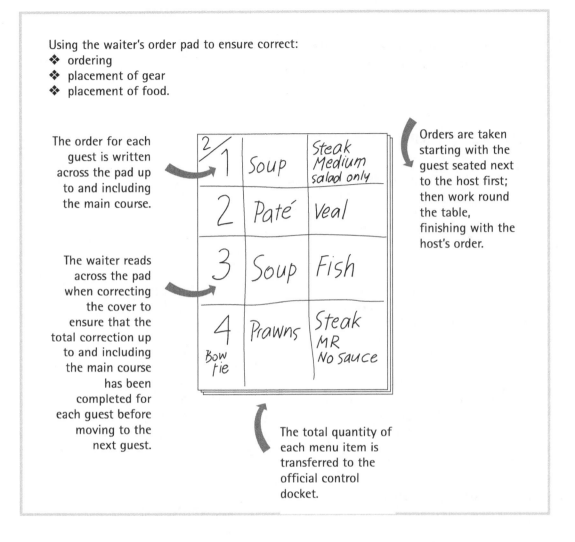

2⁄1	Soup	Steak Medium salad only
2	Paté	Veal
3	Soup	Fish
4 Bow Tie	Prawns	Steak MR No sauce

Orders are taken starting with the guest seated next to the host first; then work round the table, finishing with the host's order.

The waiter reads across the pad when correcting the cover to ensure that the total correction up to and including the main course has been completed for each guest before moving to the next guest.

The total quantity of each menu item is transferred to the official control docket.

Assuming plate service, the plated food would be placed in the waiter's left hand starting with the prawns, followed by the soup and then the pâté in a three-plate method. The remaining soup would be carried in the right hand. The plated food is now positioned to allow the waiter to place the correct plate (soup) in front of the person on the left of the host, moving and serving the second and third (pâté and soup) plates. The prawns for the host are placed last.

If the main item on the plate is turned so that it is closest to the waiter when it is picked up, then the plated meal will automatically be in the correct position and will be placed with the main item correctly placed closest to the guest.

ELECTRONIC BILLING MACHINES

Accuracy and speed are the biggest advantages of this system. Bills may be prepared by the waiter, thus cutting out the need for a cashier.

All waiters will have a key with their own letter on it, a stationery folder, food order pads and bills with consecutive numbers, a paying-in slip and a float.

When taking the order, each course is written on the order slip and placed in the billing machine before being taken to the kitchen. The correct keys must be pushed to price the order accurately. To use the machine, waiters must insert their key into the appropriate position.

How to use an electronic billing machine

1. Place the waiter's key into the correct locking position.
2. Place the docket in the machine.
3. Press buttons to price the dishes concerned.
4. Press the food/beverage button to show what the money was received for.
5. Press identity key.
6. The docket will now print out.

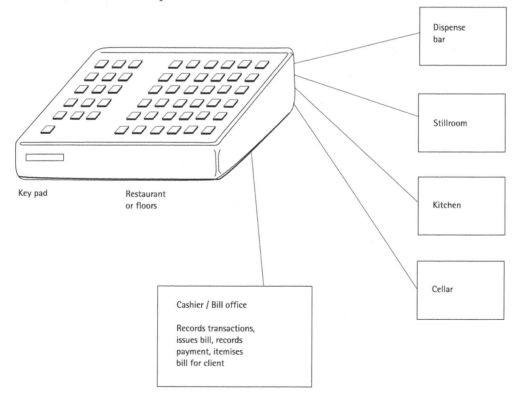

Dispense bar

Stillroom

Kitchen

Cellar

Key pad

Restaurant or floors

Cashier / Bill office

Records transactions, issues bill, records payment, itemises bill for client

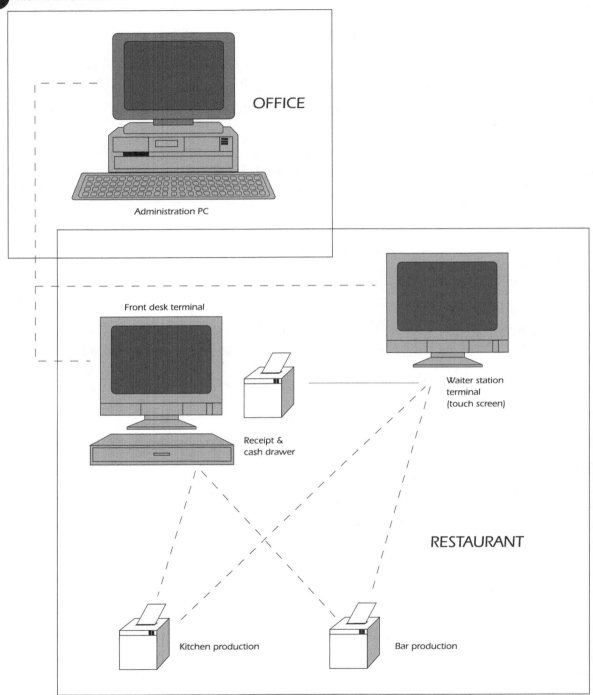

OFFICE

Administration PC

Front desk terminal

Receipt &
cash drawer

Waiter station
terminal
(touch screen)

RESTAURANT

Kitchen production

Bar production

A Touch Screen POS System (with acknowledgement to the Red-Cat Systems)

See 'Computerised control systems' on the following page.

COMPUTERISED CONTROL SYSTEMS

Computerised control systems are now widely used in restaurants, particularly in residential hotels and large establishments such as clubs. When this type of system is used, orders are entered by waiters at one or more terminals situated on the floor. Orders may be 'over-the-counter' sales or saved to allocated tables.

Production dockets are automatically printed to dispensing points, such as the kitchen or bar, and area-grouped in classes such as entrées, mains, desserts and beverages. Itemised receipts are printed for presentation to the customer on demand.

The system tracks all operations by waiter, table or sale identification. Special functions such as deleting items, clearing sales, editing paid accounts or resetting terminals are passcode-protected and access is provided to selected personnel (by management) via their personal passcodes.

Comprehensive reporting provides for sales analysis, product mix breakdowns, performance reporting by waiter, and a complete breakdown for reconciliation purposes.

In addition to offering a fast, efficient floor service, the point of sale (POS) computerised control system provides full management control and business administration, including:

❖ debtors and creditors administration

❖ full accounting functionality

❖ bank account administration and reconciliation

❖ comprehensive financial reporting

❖ hospitality-specific payroll processing

❖ recipe and function costing

❖ stock allocation and ingredient needs

❖ manual or perpetual stocktake functions

❖ automatic reordering and purchase order generation

❖ automatic calculation of stock levels and values.

Use of the service plate

A **service plate** is a dinner plate, covered with a folded napkin to reduce noise. It is used to carry all small items used in food service to and from the table—cutlery, cruets, etc.

Service plates should be readily available at all service points. At no time should waiters carry cutlery in their hands.

❖ Carry the service plate on the flat of the left hand.

❖ Use the right hand to lift the items from the service plate and place them on the table.

The service plate

Correcting the covers

To correct a cover is to adjust the cutlery originally laid to meet a guest's specific order. Covers are corrected after the orders have been taken and placed with the kitchen. At this time they are corrected up to and including the main course.

Guests use the outer cutlery for their first courses and move inwards for each succeeding course.

Cutlery for the dessert and the cheese is corrected after those orders have been taken later in the meal.

❖ Prepare the cutlery for each guest, up to and including the main course, on a service plate.

❖ Starting with the guest whose order was taken first, move around the table, correcting the covers.

❖ Correct the knife section of the first guest and the fork section of the next guest by standing between them to prevent the need to lean across the front of the guest.

Correcting the covers

❖ To adjust the cutlery, lift the item not required and replace it with the correct item. Place the items required in sequence of use; that is, with the first-course items on the outside and the items for the later courses inside and nearer the plate in the order in which they will be used.

❖ Pick up the cutlery, holding it between the thumb and index finger at the neck or join between the handle and the top of the gear (the appropriate cutlery). This will ensure that no fingerprints can be seen on the cutlery after it has been placed.

❖ All cutlery used when adjusting the covers is placed parallel to the main gear. This applies to dessert spoons and forks when the covers are adjusted for the dessert and cheese orders, as well as to the cutlery used for the earlier courses.

SIDE SALADS

Whether the guest orders a side salad or one is automatically served as an addition to a particular dish, the salad must be placed on the table so that it is convenient for the guest but doesn't over-clutter the table.

A salad should be served in bite-size pieces so that a salad fork (a small fork—see Chapter 3) is the only piece of cutlery that should be required.

The procedure is:

❖ Collect the salad forks while preparing to correct the covers for each guest at the table up to and including their main course.

❖ While correcting the fork section of the cover for each guest, place the salad fork directly to the left of, and parallel with, the main fork.

❖ It is usual to place side salads on the table just before the main course is served.

❖ Individual salads should be placed on the table from the guest's left, and positioned at the top of the fork section.

DESSERT GEAR

As we have already noted (see Chapter 4), dessert gear is usually placed or corrected after the main course has been cleared, just before the dessert course is served. This means that the dessert gear must be brought to the table and placed.

The procedure is:

❖ Arrange the required dessert gear for the whole table on a service plate and take it to the table.

❖ Correct the cover by standing between each pair of guests and placing a spoon to the right of the guest on your left and a fork to the left of the guest on your right. Move around the table repeating the procedure.

❖ If the dessert cutlery required is only a small spoon and not a full-size dessert spoon and fork, the small spoon is placed to the right of the cover.

If the dessert gear was placed on the table across the top of the cover as part of the cover for a set menu, it must be moved down before the dessert course is served. Stand between two guests and move the spoon of the guest to your left and the fork of the guest to your right, repeating the procedure around the table as you would do if you brought the dessert gear to the table on a service tray.

QUESTIONS

1. How should a menu be presented?

2. What is suggestive selling?

3. What qualities are needed for a successful salesperson?

4. Give examples of three reasons why a guest may require a special diet.

5. Name four common food allergies that people may suffer from.

6. What foods would be acceptable to a person on a Lacto diet?

7. What is the name given to a diet that is strict vegetarian?

8. The following cultural foods relate to which religious groups?

    (a)   Kosher

    (b)   Halal

    (c)   Prashad

9. What are two good ways to keep children occupied while they are waiting for their food?

10. What are the five essential points to make clear on an order?

11. List three functions normally passcode-protected from waiters in a computerised control system.

12. Why should a service plate be used for carrying cutlery?

13. What is the procedure for correcting the covers?

14. Where should an individual salad be positioned in the cover?

15. When should the dessert cutlery be corrected?

Chapter 7

Styles of service: plate service

Why should guests have to experience delays in the service of food to their table caused by a waiter's inadequate skills?

When you have completed this chapter you will have a good understanding of:

❖ Plate service procedures

❖ The use of service cloths for carrying hot plates

❖ Two-plate carrying technique

❖ Three-plate carrying technique

❖ The use of underliners.

The manager of a food service establishment has the choice of a number of different styles of service. Management should select a style of service to complement the type of food being served, taking into account the effectiveness of the service to accommodate both the guests' and the establishment's needs.

An operator who understands the various service styles can utilise a particular service, or combine two or more different styles of service, to achieve greater productivity (and profitability) and enhance guest satisfaction.

The two most commonly used food service styles are *plate service* and *silver service*. The various styles of service and the techniques required by them are all explained in detail in this or later chapters. They are:

❖ plate service (this chapter)

- ❖ silver service (Chapter 8)

- ❖ guéridon service (Chapter 10)

- ❖ family service (Chapter 10)

- ❖ smorgasbord service (Chapter 10)

- ❖ buffet service (Chapter 10)

- ❖ cafeteria service (Chapter 10).

These styles of service can be used in new and interesting ways to encourage custom. Cafeteria service, for example, is used in modern food halls, while guéridon service is used imaginatively for salads and desserts in some bistros and delis. Silver service is sometimes used instead of the usual plate service at large tables at banquets.

In plate service, the food is plated in the kitchen or at a service point and served to the guest on the plate. In silver service, the food is presented on serving dishes and is transferred from a serving dish to the plate in front of the guest with the use of a spoon and fork—the 'service gear'.

There are, of course, variations on these two basic techniques. For example, the main food item, perhaps the meat, can be placed on the plate in the kitchen and served using the plate service technique, and then the vegetables can be served at the table using the silver service technique. This constitutes a combination of the two service techniques, not a new technique.

This chapter deals with plate service techniques and procedures. Silver service technique is explained in the next chapter.

Whatever the style of service, all waiters should remember to bend the forward knee when placing items on the table. This will reduce the risk of back strain.

Plate service skills and techniques

Plate service is a basic and commonly used form of service. It requires the waiter to be skilled in carrying plates without disturbing the food arranged on them. The methods used to carry the plates depend on the number of plates to be carried.

In professional plate service, no more than four plates are carried at a time. It is possible to carry more than four plates but, as this relies on balance, it is not usually considered professional service.

The two professional methods most utilised in the industry are the two- and three-plate carrying techniques. These involve carrying either two or three plates in the left hand, leaving the right hand free. The right hand can be used to carry another plate, thus allowing three or four plates to be carried at once.

When plates are cleared, the same plate-carrying techniques should be used (see Chapter 9).

All waiters must be proficient in plate carrying and clearing techniques.

Plate service procedure

Traditional plate service required food to be served from the left of the guest, and empty plates cleared from the right. In modern plate service, however, plates are both placed and cleared from the guest's right, as this causes the least disturbance to the guest.

Modern plate service practice developed because dining space is now more intensively utilised than in the past, and there is less room for movement between guests. The plate service waiter can unobtrusively place a plate of food in front of a guest from the right, while holding other plates in the left hand safely behind the guest's head. Left-handed waiters may reverse the technique and serve and clear from the left.

In modern plate service, plates are placed from the guest's right

The modern plate service practice of both placing the plates and clearing them from the guest's right has been adopted by leading training institutes and establishments throughout the world. However, there are restaurants that still offer traditional plate service from the left. Waiters must, of course, conform to the 'house rule' on this point.

Modern plate service doesn't interfere with beverage service, as food and beverage service don't take place at the same time.

Unless otherwise instructed, first serve the guest seated next to the host and then move around the table serving each guest in turn, regardless of sex. The host should be served last. Note, though, that in some establishments you may be required to serve ladies before gentlemen, or this may be requested by the guests.

THE SERVICE CLOTH

Service cloths should be used to protect the hands and wrists from burns when you are carrying hot plates.

❖ Place the cloth along your left hand and forearm, with the open section of the cloth inwards; the cloth should not protrude beyond the tips of the fingers.

❖ Fan the end of the cloth open to protect the hand: this allows both the two-plate and three-plate carrying methods to be used (see below).

❖ A second folded service cloth should be held in the right hand to protect it when holding plates or transferring them from the left hand to the table.

Using a service cloth to carry hot plates

Two-plate carrying technique

❖ Plates are picked up so that, when they are placed, the main item will be on the side of the plate facing the guest. Remember that the first plate to be picked up in the left hand will be the last to be placed on the table.

❖ Hold the first plate between your thumb, index finger and middle finger of your left hand. If the plate is hot, use a service cloth (see above).

❖ Then place the second plate on a platform above the first plate, supporting it by your ring (or fourth) finger, your little finger, and the base of your thumb and lower forearm.

❖ You may carry a third plate in your right hand, also in a service cloth.

❖ Carry the plates to the table, holding them away from your body, with your shoulders held back, so that the plates are not resting against the front of your body.

❖ Only bring the plates in front of your body when limited access to the table requires it.

❖ To place the plates in front of the guests, position yourself at the back right-hand corner of a guest's chair, holding your left hand (and its plates) out of the way behind the guest's head.

❖ Step forward and place the plate in your right hand in front of the guest from the guest's right. (Remember to bend your forward knee as a precaution against back strain.)

❖ The plate should be placed so that the main item (the meat, fish, etc.) is immediately in front of the guest, and the vegetables further away at the 'top' of the plate.

❖ Move behind the next guest and transfer the second plate to your right hand and place it in front of the guest from the guest's right.

❖ Continue around the table repeating the procedure.

Two-plate carrying technique

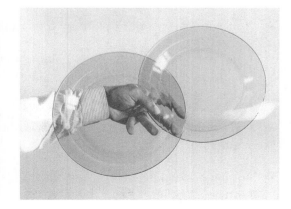

Two-plate carrying technique

Three-plate carrying technique

If four plates must be taken to the table at the same time, three plates should be carried in the left hand using the three-plate carrying method.

❖ Hold the first plate between your thumb, index finger and middle finger of the left hand (as in the two-plate technique). If the plates are hot, use a service cloth (see above).

❖ Place the second plate into the crease of the palm of your left hand under the edge of the first plate, supporting it by your ring finger and little finger.

❖ Place the third plate so that it sits on the flat of your forearm and the rim of the second (lower) plate.

Three-plate carrying technique

❖ Carry the fourth plate in your right hand.

❖ Carry the plates to the table, holding them away from your body, with your shoulders held back, so that the plates are not resting against the front of your body.

❖ The plates should only be carried in front of your body when limited access to the table requires it.

❖ To place the plates in front of the guests, position yourself at the back right-hand corner of the guest's chair, holding your left hand and its plates out of the way behind the guest's head.

❖ Step forward and place the plate in your right hand in front of the guest from the guest's right.

❖ Move behind the next guest and transfer the third plate to your right hand and place it in front of the guest from the guest's right.

❖ Continue around the table repeating the procedure.

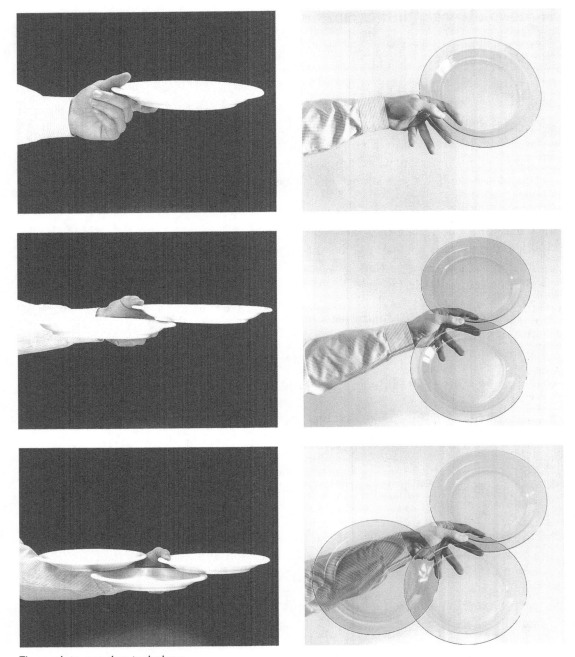

Three-plate carrying technique

Underliners

An **underliner,** or underplate, is a plate placed under the service equipment (the vessel containing the food) when it is served to guests. Underliners are not primarily intended to enhance the appearance of the food being served; their purpose is to make it easier to carry and clear service equipment that is difficult to handle, and to provide something for guests to place their used cutlery on when the service equipment itself isn't suitable for that purpose.

It is usual for a napkin or doily to be placed on the underliner to prevent the service equipment from sliding.

Underliners should only be used for a specific purpose. They are often used when serving items in soup bowls and coupes. They should also be used when serving natural oysters on ice to prevent condensation.

Underliners used for soup bowls and coupes

QUESTIONS

1. Which are the most commonly used food service styles?

2. What is the maximum number of plates that a professional waiter should carry to the table at once?

3. On which side of the guest does the waiter stand in modern plate service?

4. What is the use of the service cloth?

5. Describe the two-plate carrying technique.

6. Where should the plates be held when carrying them to the table?

7. Where should the main item be on the plate when it is placed in front of the guest?

8. Describe the three-plate carrying technique.

9. What are underliners used for?

10. How can you stop the plates sliding around on their underliners?

11. How can a waiter reduce the risk of back strain when placing items on the table?

Chapter 8

Styles of service: silver service

Obviously, this waiter hasn't mastered the basic silver service skills—he's using kitchen tongs instead of a spoon and fork!

When you have completed this chapter you will have a good understanding of:

❖ The technique of using a serving spoon and fork

❖ The technique of using knives for service

❖ Silver service of bread rolls

❖ Silver service procedures for the main meal

❖ Silver service of sauces

❖ Silver service of delicate or large items.

Silver service is the technique of transferring food from a service dish to the guest's plate from the left with the use of service gear. (If silver service was performed from the right, the service plate would present a barrier between the guest's plate and the waiter's right hand used for transferring the food items.) **Service gear** usually means a serving spoon and fork, but occasionally it may consist of knives, especially fish knives. Silver service requires the waiter to be able to use the service gear to serve the food with one hand. It is essential that silver service is performed with well-practised technique so that the food remains hot and the guests are served promptly.

Silver service takes place from the left of the guest

Professional silver service depends on mastering the technique of using service gear held in one hand to transfer items to the guest's plate from a service dish held in the other hand.

Use of a spoon and fork (wedge method)

❖ Place a fork over a spoon in your right hand, both facing up. They should rest across your middle, ring and little fingers, with the base of the gear resting in line with the bottom of the little finger, leaving the index finger and thumb free to move the gear. The handles shouldn't protrude beyond your little finger.

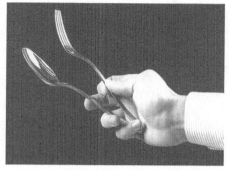

Holding the service gear

❖ Slide your index finger between the fork and the spoon, and hold the fork between the tip of your index finger and the tip of your thumb so that you can lift the fork with your index finger and thumb.

❖ Holding the fork between the tips of your index finger and your thumb, raise the fork from the bowl of the spoon, keeping the ends of the handles of the fork and spoon together with your little finger.

For large round items turn the fork

❖ At no time allow the index finger and thumb to slide more than halfway up the handles of the gear.

❖ You can now lift food items with the spoon and hold them firmly in place with the fork while you transfer them to the guest's plate.

❖ Pick the food up from the side, drawing the gear towards you as you lift it.

❖ If the item to be moved is small or very thin, you can remove your index finger. This enables the fork to be pressed more tightly against the spoon, holding the item firmly. To release the item, insert the tip of your index finger to separate the gear.

❖ For large round items, such as bread rolls, you may turn the fork, enclosing the item.

❖ The tops of some items—broccoli hollandaise, for example—must not be touched when they are served. In such cases, turn your hand so that you can pick up the item from its sides.

Some waiters thread the fork through their fingers when using the service gear for silver service (thread method). To the novice this may seem an easier and safer technique, but it has its limitations. It is much too slow, for example, when several different items requiring different service gear have to be served quickly one after the other. Also, it is difficult to apply when two knives are being used as the service gear (see below).

Use of knives for service

Two knives, preferably fish knives, are sometimes used for the service of soft or large items requiring more support than can be given by a service spoon and fork.

❖ Hold the two knives in the same fashion as a spoon and fork.

❖ Fan out the knives to give greater support to the item you will be moving.

❖ Place the knives under the food item, supporting it while you transfer it to the guest's plate.

Silver service of bread rolls

❖ Hold the bread basket on the flat of your left hand.

❖ Silver service takes place from the left of the guest.

❖ Place your left hand low above the edge of the guest's side plate (no more than 5cm above the plate).

❖ Hold the bread roll in the service gear and transfer it to the side plate.

❖ Move around the table offering a bread roll to each guest.

Silver service procedure: the main meal

❖ Place clean hot plates in front of the guests from the right from a stack carried in the left hand.

❖ Hold the serving dish on a service cloth on the flat of your left hand, with your hand under the centre of the service dish.

Place clean hot plates in front of each guest

❖ Calculate the size of the portion allowed for each guest and decide how it should be presented, considering the other items to be served. (For example, two green vegetables may be separated by a white item, giving a satisfactory visual balance.)

❖ As noted above, silver service takes place from the left of the guest.

❖ Hold the serving dish over the guest's plate, no more than 5cm above it.

❖ Place the main item of the course to the front of the guest's plate.

❖ Place the vegetables around and behind the main item away from the guest in the pattern already decided for plate presentation.

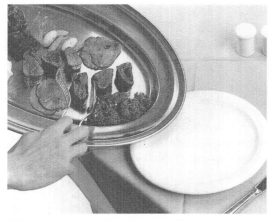

When a food item is picked up from the serving dish the service gear is moving towards you

When the item is released on the guest's plate the service gear is moving away from you

❖ Note that no items are placed on the rim of the plate.

❖ Garnishes are placed to enhance the presentation.

❖ If you are serving small items such as button mushrooms, transfer a small number at a time. You can then position them carefully on the plate.

❖ Pick up food items on the side of the serving spoon closest to you (its left side) and release them from the same side of the spoon. When a food item is being picked up from the serving dish, the service gear is moving towards you; when the item is released on the guest's plate, the service gear is moving away from you.

❖ Plate presentation should be consistent: the different food items in a dish should always be presented in the same pattern to all guests who have selected that dish.

❖ Move around the table serving each guest in turn, with the host last.

Silver service of sauces

In silver service, accompanying sauces should not be carried on the same serving dish as the food. They are offered and served separately, using a sauce-boat (*saucière*) and a serving spoon or special ladle. Sauces are never poured direct from the sauce-boat. A spoon (or ladle) is always used to serve them.

❖ Carry the sauce-boat on an underplate on the flat of your left hand, with the lip of the sauce-boat facing to the right.

❖ Serve the sauce from the left of the guest.

❖ Lower the underplate over the guest's plate, so that it is not more than 5cm above it.

❖ Hold the serving spoon (or ladle) in your right hand with the handle of the spoon above the lip of the sauce-boat.

❖ Draw the spoon across the sauceboat towards you to collect the sauce.

❖ Carry the spoon away from you to sauce the appropriate item. Note that the sauce should cover only one-third of the item.

Draw the spoon across the sauce-boat towards you to collect the sauce.

❖ Other accompaniment items, such as mustard or apple sauce, are placed to the left of the main item.

❖ Move around the table offering the sauce or accompaniment to each guest in turn, with the host last.

Silver service of sauces

Silver service of delicate or large items

Two knives, preferably fish knives, are used to serve delicate or large items that are difficult to handle with a service spoon and fork.

❖ Two (fish) knives take the place of the service spoon and fork.

❖ Hold the knives in your right hand in the same way as a service spoon and fork, and fan them out (see above).

❖ Transfer the food item to the guest's plate with the fanned knives.

❖ Remove the knives when the food touches the plate.

❖ In addition to the knives, carry a set of ordinary service gear (a serving spoon and fork) on the serving dish to serve the vegetables and garnishes.

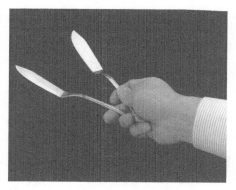

Hold the knives in your right hand and fan them out

QUESTIONS

1. What is silver service?

2. Which side of the guest do you stand for silver service?

3. What are the limitations of using the thread method of holding the service gear?

4. How do you pick up small items of food?

5. What sort of things are served with two fish knives instead of the usual service gear?

6. In which direction should the gear be moving as you pick up food from the service plate?

7. How would you plan the presentation of the food on the plate?

8. How and when are sauces served in silver service?

9. Where are other accompaniments placed?

Chapter 9

Clearing the table

Is this the same waiter who has been so helpful and knowledgeable?

On completing this chapter you will have a good understanding of:

❖ Clearing procedures

❖ Two-plate clearing technique

❖ Three-plate clearing technique

❖ Clearing side plates

❖ Clearing soup bowls and oddly shaped dishes.

The same clearing techniques are used for both plate service and silver service. They rely on plate-carrying techniques similar to the two-plate and three-plate techniques already described (see Chapter 7).

When a course is cleared, it is usual for the whole table to be cleared at the same time when all the guests have finished. Guests usually indicate that they have finished by placing their cutlery together on the plate. As they don't always do this, you must be alert to other signs from the table that everyone has finished; if you are doubtful whether guests have finished or not, you should ask them.

Once you have established that all the guests have finished the course, clear the plates using one of the following techniques.

Two-plate clearing method

❖ Start with the person sitting next to the host.

❖ Standing at the back right-hand corner of the guest's chair, lean forward (bending your right knee, as you should when placing a dish in front of a guest) and pick up the used plate and cutlery with your right hand.

❖ Transfer the plate to your left hand, holding it between the thumb and index finger. Place your thumb over the end of the fork handle. Use the knife to move the scrap items to the front of the plate.

Clearing the table – these guests have indicated that they have finished by placing their knives and forks together on their plates

❖ Place the knife under the handle of the fork at right angles to it.

❖ Moving around the table, place yourself behind the next guest. Holding your left hand (and the first guest's empty plate) behind the guest, lean forward and pick up the second used plate and its cutlery.

❖ Transfer the second plate to your left hand. Position it on a platform above the first plate, supporting it with your ring finger, little finger and the base of your thumb and lower forearm.

❖ Place the fork alongside the other fork on the first plate and, using the knife, push the scraps down off the second plate on to the front of the first plate to join the scraps already there.

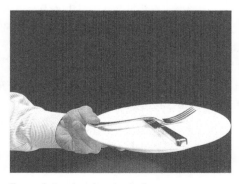

Two-plate clearing technique

❖ Place the knife alongside the knife on the first plate.

❖ Moving around the table, collect the remaining plates and cutlery. Stack the plates on the second plate and arrange the cutlery on the first plate, following the same procedure as for the second plate.

❖ The number of plates that can be collected in this way will depend on the waiter's skill and experience. When you have collected as many plates as you can confidently carry, take the plates and cutlery to the station (sideboard) and place them on a tray for removal, or take them directly to the dishwashing area, according to the practice of the establishment.

Three–plate clearing method

The three-plate clearing method is similar to the two-plate method, with the added advantage that the scrap food items and the used cutlery are carried on separate plates.

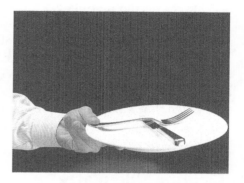

❖ Start with the guest sitting next to the host.

❖ Standing at the back right-hand corner of the guest's chair, lean forward and pick up the used plate and cutlery with your right hand.

❖ Transfer the plate to your left hand, holding it between the thumb and index finger. Place your thumb over the end of the fork handle. Use the knife to move the scrap items to the front of the plate.

❖ Place the knife under the handle of the fork at right angles to it.

❖ Moving around the table, place yourself behind the next guest. Holding your left hand (and the first guest's empty plate) behind the guest, lean forward and pick up the second used plate and its cutlery. (Up to this point, the technique has been exactly the same as for the two-plate method.)

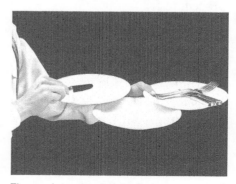

Three-plate clearing technique

❖ Place the second plate in the crease of the palm of your left hand under the edge of the first plate, supporting it by your ring finger and little finger. Place the fork alongside the fork on the first plate and, using the knife, move the scrap items from the first plate down on to the second plate. Place the knife alongside the knife on the first plate.

❖ Moving around the table, pick up the next guest's used plate.

❖ Place the third plate so that it sits on the flat of your forearm and the rim of the second plate. Place the fork alongside the forks on the first plate and use the knife to move the scraps on to the second plate. Place the knife alongside the other knives on the first plate.

❖ Continue collecting the plates, stacking the additional plates on the third plate, transferring the scraps on to the second plate and placing the knives and forks neatly on the first plate.

❖ When you have collected as many plates as you can confidently manage, take them to the station (or the dishwashing area).

Clearing side plates

CLEARING SIDE PLATES AT THE SAME TIME AS DINNER PLATES

If there are only three or four guests at the table, the side plates may be collected at the same time as the used dinner plates, using the two-plate or the three-plate technique. Continue around the table a second time, collecting the side plates and knives. If using the two-plate technique the procedure is:

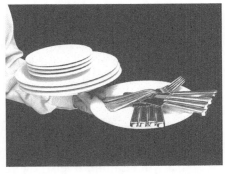

Clearing side plates with dinner plates: two-plate technique

❖ Collect the side plates and knives from the guests' left using your right hand.

❖ Transfer the side plate to the pile of empty plates supported by your left hand and arm, holding your left hand well away from the table.

❖ Use the knife to move scrap items to the front of the first dinner plate.

Three-plate technique

❖ Place the knife on the first plate beside the other knives.

❖ Continue until all the side plates have been collected and stacked on the dinner plates.

If the three-plate clearing technique is used, collect the knives and forks on the first plate and the scraps on the second plate, piling the side plates on the cleared main course plates, which are stacked on the third plate position.

CLEARING SIDE PLATES SEPARATELY FROM DINNER PLATES

If there are more than four guests at the table, you will not be able to collect the side plates at the same time as the dinner plates. Collect them separately, using the two-plate method.

❖ Take a dinner plate to the table. It will provide you with a conveniently larger working surface than a side plate.

❖ Treat this plate as if it were the first dinner plate collected, and use it as the receptacle for scraps and the side knives.

Clearing side plates separately from dinner plates

❖ Standing to the left of the guest, lean forward and pick up the used plate and cutlery with your right hand.

❖ Transfer the plate to your left hand, positioning it in the second plate position (two-plate method). Transfer scrap items to the first plate and position the first side knife with the handle secured under your thumb.

❖ Moving around the table, place yourself behind the next guest. Holding your left hand (and the first guest's empty side plate) behind the guest, lean forward and pick up the second used plate and its cutlery.

❖ Continue around the table until all the side plates are cleared, positioning the additional knives under the first knife.

Soup bowls, coupes and odd-shaped serving dishes

Untypical and relatively difficult items such as soup bowls, coupes and oval pasta dishes will usually have been served on an underliner (see Chapter 7). They should not be stacked but held separately from the used dishes, using the two-plate or three-plate carrying method.

Clearing odd-shaped items

QUESTIONS

1. How can you tell when all the guests have finished a course?

2. Describe the two-plate clearing technique.

3. Where should the dirty plates be positioned in the hand when removed from the table?

4. How is the cutlery prevented from falling off the plate as it piles up?

5. Describe the three-plate clearing technique.

6. What is the advantage of the three-plate clearing technique?

7. How many plates should you collect at once?

8. What technique is used to collect side plates at the same time as dinner plates?

9. When should side plates be collected separately from dinner plates?

10. How should unusually shaped plates be collected?

Chapter 10

Other forms of service

No wonder the guests get confused at a buffet or smorgasboard when the staff don't understand what they should be offering.

On completion of this chapter you will have a basic understanding of the techniques and procedures of:

❖ Soup service from a tureen or individual soup cup

❖ Guéridon service

❖ Family service

❖ Smorgasbord service

❖ Buffet service

❖ Cafeteria service.

Apart from the basic table service procedures—plate service and silver service—a waiter may encounter and be expected to be proficient in other specialist forms of service. Some of these are described in this chapter.

Soup service from a tureen or soup cup

A tureen is a deep bowl with a lid to keep the contents warm. Individual soup cups or, for a number of people, one large tureen may be used as an alternative to the plate service of (pre-filled) soup bowls. Individual soup cups (occasionally with covers) may be offered in a silver service procedure, while service from a large tureen is best performed at a guéridon.

The use of a tureen ensures that the soup stays hot until it is served at table.

INDIVIDUAL SOUP CUPS

❖ Carry a full individual soup cup on a service plate held on the flat of your left hand to the guest.

❖ Pour the soup from the cup into the guest's soup bowl, pouring away from the guest.

❖ If the soup contains items that cannot easily be poured, such as vegetables or noodles, carry a dessert spoon on the service plate and use it to spoon them into the soup.

LARGE TUREENS

❖ Set a guéridon with hot soup bowls, underliners and a soup ladle, and place it near the table (see below).

❖ Take the lid off the tureen and present the soup to the host.

❖ When the soup has been approved, place the open tureen on the guéridon.

❖ Hold a soup bowl on an underliner in the left hand, and use the right hand to ladle soup from the tureen into the bowl.

❖ Scoop the soup from the bottom of the tureen to ensure an even consistency.

❖ Place the soup bowls in front of the guests from each guest's right, using the usual plate service procedure.

❖ If the soup requires extra garnishing, serve the garnish from the left of the guest, using silver service procedure.

Guéridon service

The term *guéridon* means a trolley (or side table) used for the service or preparation of foods in the dining environment.

Guéridon service specifically refers to the transfer of food from a serving dish to the plate on a guéridon. Guéridon *preparation* is the preparing or finishing of foods on the guéridon—for example, table-cooking or the tossing of salads.

This section is concerned with guéridon service only. Guéridon preparation is dealt with in Chapter 12.

GUÉRIDON SERVICE TECHNIQUE

❖ Set the guéridon with the appropriate mise-en-place for the service or preparation to be performed. It is essential that the mise-en-place is complete with all items before service begins. You should not leave the guéridon during service.

Hold the spoon below the fork as you collect the food

❖ Place clean hot plates on the guéridon.

❖ Present serving dishes containing the food prepared in the kitchen to the guests, and then place them on the guéridon.

❖ At the guéridon use both hands to manage the service gear, holding the serving spoon in your right hand and the fork in your left.

❖ Hold the spoon below the fork as you collect the food.

❖ Position the main item to the front of the plate, with the vegetables around and behind it, allowing some space between the different food items, as in silver service (see Chapter 8).

❖ Sauces may be served at the guéridon or at the table, using the silver service technique.

❖ Place the plated meals in front of the guests from their right, using the usual plate service technique.

Family service

Family service is a simple method of service in which serving dishes are placed on the dining table, allowing the guests to select and serve themselves. This style of service enables the guests to select only what they require and in appropriate portions.

Family service is often offered in addition to plate service. For example, the main item may be plate-served and the guests left to help themselves to vegetables or salad.

FAMILY SERVICE TECHNIQUE

❖ Before serving plates are placed on the table, you must make room on the table for them.

❖ Using a service cloth, place clean hot plates from a stack in front of the guests from the guests' right.

❖ Place the serving dishes on the table, each with a set of serving gear resting on the sides of the dishes.

❖ The serving gear should be placed so that the handles face the nearest guest.

❖ The serving dishes may be removed when they are empty at any time during the meal.

Smorgasbord service

In **smorgasbord** service, guests select from a presentation of food items, hot or cold, serving themselves directly on to their plates without the help of service staff.

It is usual practice for the guests to select only one course at a time, returning to the smorgasbord to help themselves to later courses.

The waiter should maintain the arrangement of the food on the platters throughout service to ensure that the food is always attractively presented.

Buffet service

In **buffet** service, as in smorgasbord service, guests move to the buffet and select what they want from a presentation of food items, hot or cold. The difference is that, in buffet service, staff serve the guests with the food they have selected, whereas in smorgasbord service the guests help themselves.

Service staff positioned behind the buffet assist the guests by plating their food for them as they select it.

Service staff use silver service technique to plate the guests' food. Hold the guest's clean plate in your left hand and, using a serving spoon and fork held in your right hand, transfer the food items selected from the service plates to the guest's plate.

As in silver service at table, you must place the selected items on the plate carefully so that they are visually well balanced and convenient for the guest.

Cafeteria service

In cafeteria service, guests collect their own meals on a tray as they select food items from the race. Modern food-hall operations often use this basic service technique. This style of service is also used in some cafés and coffee shops.

Buffet, smorgasbord and cafeteria waiting staff, like all other waiters, must have a good knowledge of the items they are serving.

Clearing and wiping down

Whatever the style of service, it is the responsibility of the waiting staff to clear used items from the table. If tables are clothed, soiled cloths must be changed. If the tables are not clothed, they must be wiped down and sanitised.

QUESTIONS

1. What is guéridon service?

2. What is a tureen?

3. How is soup served from a large tureen?

4. What is the difference between guéridon service and guéridon preparation?

5. Describe guéridon service technique.

6. How are the sauces served in guéridon service?

7. What is family service?

8. When should the serving dishes be removed during family service?

9. What is the difference between smorgasbord service and buffet service?

10. What is cafeteria service?

Chapter 11

Other food service procedures

The little touches make the difference.

On completion of this chapter you will have a basic understanding of how to:

❖ Crumb (or brush) down the table

❖ Serve hot or cold towels

❖ Change ashtrays.

Crumbing/brushing down

Tables are usually crumbed down after the main course and side plates have been cleared.

Although a variety of crumbing implements, such as brush and pan sets, table scrapers and electric brushes, is available for this purpose, the most commonly used equipment is a dinner plate and a folded service cloth. This basic equipment is, of course, readily available in all styles of establishment.

CRUMBING DOWN USING A SERVICE PLATE AND CLOTH

❖ Make sure the side plates, cruets and other items no longer required have been removed.

❖ Hold the plate on the flat of your left hand with your hand under the centre of the plate.

❖ Brush down from the guest's left (from where the side plate was before it was cleared).

❖ Hold the plate just under the edge of the table with your left hand.

❖ Brush the crumbs on to the plate using a folded service cloth held in your right hand.

❖ Don't flick the crumbs, but brush them steadily towards you with the folded service cloth.

❖ Move around the table crumbing down each guest's place as required, finishing with the host.

Service of hot or cold towels

Crumbing down

Hot or cold towels have been offered for many years on airlines and in establishments with Asian cultural values. Because guests came to appreciate this so much, hospitality venues of all kinds now sometimes offer the service. Hot or cold towels or flannels are offered to the guests so they can refresh themselves either before or after the meal. The towels are moistened and lightly sprayed with a cologne.

Whether the towels are hot or cold is likely to depend on the climatic conditions. Moistened and slightly scented towels, either hot or cold, provide a refreshing start or finish to the dining experience.

Offering a cold towel using tongs

THE TECHNIQUE OF SERVING TOWELS

❖ Stack the towels (rolled) on a serving tray together with a set of tongs or service gear for service.

❖ Carry the tray of towels on the flat of your left hand.

❖ Offer the towels from the guest's right.

❖ Holding the tongs or service gear in your right hand, use them to grip the edge of the towel so that it will open as you offer it to the guest.

❖ Move around the table offering a towel to each guest, with the host last.

❖ Place the serving tray in the centre of the table so that the guests can put their towels on it after they have finished with them.

❖ Remove the tray with the used towels on it.

Ashtrays

Although smoking is no longer permitted indoors, there are still some external areas where smoking is allowed.

It is the responsibility of the waiting staff to ensure that used ashtrays are removed and cleaned frequently, and to see that clean ashtrays are always available to smokers.

CHANGING ASHTRAYS

When clearing a used ashtray it must first be covered to avoid the possibility of spilling ash. When the dirty ashtray has been covered and removed, replace it immediately with a clean one.

❖ Hold a clean ashtray in your right hand and place it over the dirty ashtray. The clean ashtray may be either way up.

❖ Remove the two ashtrays together using the one hand.

❖ Transfer the dirty ashtray to your left hand.

❖ Use your right hand to place a clean ashtray the right way up where the dirty one was.

Changing ashtrays

Dirty ashtrays should be held in the left hand or on a service tray well away from the guests.

QUESTIONS

1. When are tables usually crumbed down?

2. How is a table crumbed down using a service cloth and plate?

3. What is the technique used to serve towels?

4. How can an ashtray be changed without any danger of the contents spilling on the table?

Chapter 12

Use of the guéridon

Is there any reason why the term guéridon should suggest 'big bucks' to guests? Or why the term should suggest a lot of extra work and extraordinary skills to waiters?

In fact, the guéridon can be used effectively in a simple way and without much expense to give guests a 'visual' of the fare. A guéridon allows waiters to demonstrate greater versatility in their service style.

On completion of this chapter you will have a basic understanding of:

❖ The definition of the guéridon and the uses to which it is put

❖ The use of the guéridon as an aid to making sales

❖ The selection of food for the guéridon

❖ Guéridon mise-en-place

❖ Guéridon food preparation techniques—tossing, cooking, boning, carving

❖ Safety in guéridon cooking

❖ Guéridon preparation of liqueur coffees.

Most people imagine that the term 'guéridon' refers to table-cooking as demonstrated in fine dining-rooms. In fact, the guéridon is simply a piece of equipment on which we serve or prepare food in the dining environment.

In this chapter, we show the basic uses of the guéridon as a simple means of offering interesting services to guests and increasing profitability to the establishment.

The guéridon and its uses

A guéridon is simply a trolley or side table used for the service or preparation of food or beverages in the dining area.

A guéridon can be an elaborate piece of movable furniture on castors, made from exquisite timbers and provided with expensive built-in cooking equipment and silver fittings. Or it can simply be an ordinary small dining table.

As we have seen, guéridon *service* means no more than the transfer of food from a serving dish to a plate on a side table or trolley in the presence of the guests (see Chapter 10).

Guéridon *preparation* is the preparing or finishing of food on the guéridon.

The guéridon as a sales aid

The guéridon can be used most effectively as an aid to sales. The display of food on a guéridon and the preparation of food in the presence of guests, correctly used, can be powerful generators of extra sales. At its simplest, the guéridon can be used to display or to prepare:

❖ hors d'oeuvres

❖ salads

❖ bakery items

❖ desserts

❖ fresh fruit.

Food selection and presentation

When selecting food for display or preparation on the guéridon, remember that the food-stuffs chosen should:

❖ be fresh and visually attractive

❖ not discolour or break down easily

❖ be simple to prepare and require a minimum of cooking time

❖ maintain their quality in spite of the limited preparation or cooking time

❖ not produce unpleasant smells or other offensive effects when they are prepared or cooked.

SALADS

Here are some key points to consider when preparing salads on the guéridon:

❖ Thoroughly clean salad items and drain them in the kitchen before they are brought to the guéridon.

❖ Prepare items in bite-size portions.

❖ Keep items chilled until they are needed for use.

❖ The salad dressing may be prepared either in the kitchen or on the guéridon.

❖ Salad items and the dressing should be tossed at the guéridon at the time of service.

MEAT ITEMS

Key points to note when selecting and preparing meat items for the guéridon are:

❖ Select a product of premium quality.

❖ Remove all fat and sinews.

❖ Trim portions, allowing for one-minute cooking time.

❖ Keep prepared items chilled, ready for use.

❖ Seal meat items to maintain the juices.

❖ Complete the cooking in the minimum amount of time.

FRUIT

When preparing fruit for the guéridon:

❖ Clean and portion the fruit.

❖ Take special care to keep fruits that may become discoloured when peeled or cut in an appropriate juice to discourage discoloration.

❖ Keep prepared fruit chilled until it is needed for use.

❖ Avoid cooking fruit beyond the point at which it breaks down, taking special care of berries and soft fruit, which break down easily.

Guéridon mise–en–place

To avoid any delay or confusion when preparing dishes at the guéridon in front of guests, it is essential that the preparation (mise-en-place) of the guéridon should be complete.

The mise-en-place includes both the equipment used to prepare the food or beverages and the equipment needed to serve them, as well as the products required to create them.

Variation of recipes

The recipes used in cooking at the guéridon may vary in detail from one establishment to another, but the essentials of the classic recipes must be maintained to ensure that they are clearly identifiable to the guest. For example, a Caesar salad always includes lettuce, eggs, anchovies, crisp diced bacon and croûtons, but parmesan cheese is optional, and the eggs may or may not be cooked.

New dishes can be created for preparation at the guéridon to suit the establishment's menu, but they must follow the rules for the selection and preparation of foods.

Guéridon food preparation techniques

The following food preparation procedures are appropriate for use at the guéridon. They will enhance guests' enjoyment of their meals and tempt others to try the same items.

TOSSING

Salads, seafoods and fruit are often suitable for tossing in front of the guests, as they can be simply prepared and will tempt others to order some for themselves.

The main ingredient may be displayed in its natural state and then tossed with previously prepared sauces, dressings or liqueurs.

If the sauces or dressings are prepared at the guéridon, their preparation must be completed before the main item is added and tossed.

COOKING

Major considerations when cooking on the guéridon are:

❖ choosing the appropriate cooking medium—for example, clarified butter or cooking oil

❖ the length of cooking time

❖ the ability to create the appropriate flavours in a minimum amount of time.

Most items cooked on the guéridon lend themselves to the additional effect of flaming (*flambé* work) (see below).

Table-cooking no longer requires expensive cooking lamps, or guéridons with inbuilt cooking equipment, as a range of good but inexpensive table-top gas-cylinder cookers is now available.

FLAMBÉ WORK

Flambé work involves lighting liquor (usually a spirit or liqueur) in a pan at the guéridon.

The procedure is:

❖ Light the cooking lamp.

❖ Pour the required quantity of the liquor into the pan.

❖ Cover the lamp's flame completely with the pan.

Pour liquor into the pan

Tilt the pan away from you

❖ Leave the liquor to warm in the pan while you place the bottle of liquor well away from the flame.

❖ Move the pan back towards you, tilting it *away* from you until the liquor just comes into contact with the flame and ignites.

❖ As soon as the liquor is alight, lift the pan slightly and move it gently in a circular motion so that the flames move around the pan.

❖ When igniting the liquor, hold your body upright. Don't bend your head and shoulders over the pan—flames can flare up unexpectedly.

Safety in guéridon cooking

Cooking at the guéridon, particularly flambé work, obviously has its dangers because it involves naked flames in the dining-room, and fuel for those flames. Special precautions are therefore necessary.

❖ Regular checks of the cooking lamps and gas bottles will ensure that potential faults—in particular, loose fittings or leaks—will be detected before they become dangerous.

❖ A fire blanket and a small hand-held extinguisher should be kept easily accessible, in a place well known to all the waiting staff.

❖ When preparing to use the guéridon, position it at a sufficient distance from the guests' table to ensure that there can be no danger to the guests. The guéridon should be at least its own width away from the table. This has the added advantage that service can take place around the table without the guéridon getting in the way of the waiter.

MAINTENANCE OF COOKING LAMPS

❖ Lamps should be stripped down and all parts cleaned regularly.

❖ After cleaning, always check that lamps work as soon as they are reassembled.

❖ Gas bottles must never be replaced near a naked flame.

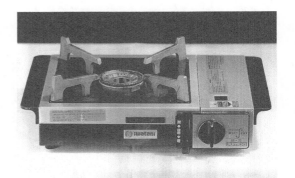

An inexpensive modern cooking lamp

❖ Always check bottles for leaks.

❖ Untrained staff must never use a cooking lamp.

Boning

If there is a side table or trolley, the waiter can use it for deboning items for guests in their presence. Fine dining rooms, employing staff with the necessary skills, are often able to offer a range of menu items requiring this form of guéridon technique.

Guéridon boning technique is most commonly used when guests are offered the service of having their (already cooked) fish deboned for them (see pages 118–119).

Carving

The carving of the joint, poultry or game at the table was for centuries a traditional part of the dining experience. Nowadays, in restaurants when the carving is done in the presence of the guests, it is usually performed at a special carvery or, where there is one, on an elaborate beef trolley, rather than on an ordinary guéridon.

Carving trolley

Liqueur coffee

Liqueur coffees are often prepared in front of guests on a guéridon because the process is visually interesting and provides an exciting finale to the dining experience. The technique of preparing liqueur coffees is explained later (see Chapter 15).

Add aerated pouring cream

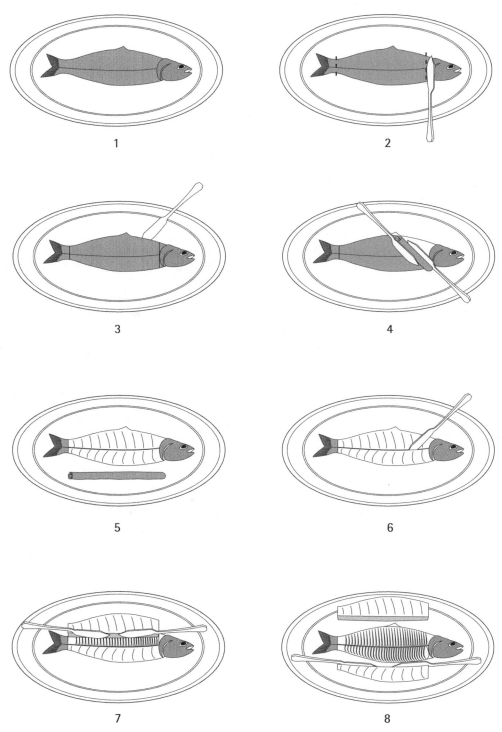

1

2

3

4

5

6

7

8

Boning fish

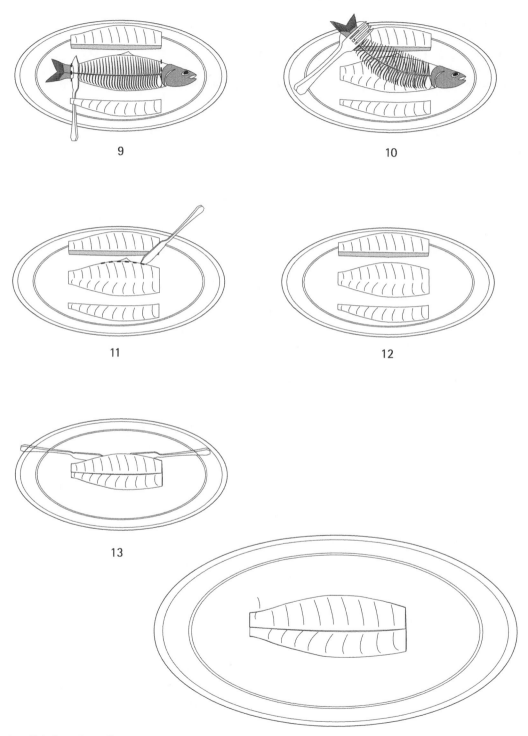

9

10

11

12

13

Boning fish (continued)

14

QUESTIONS

1. How can use of the guéridon help sales?

2. What five considerations should be borne in mind when selecting food for preparation on the guéridon?

3. At what point should salad be tossed if it is done at the guéridon?

4. Why do meat portions have to be small for cooking at the guéridon?

5. How can you prevent fruit from becoming discoloured?

6. Why is it especially important that the mise-en-place is complete before cooking on the guéridon begins?

7. What is the procedure for tossing a salad at the guéridon?

8. What is the correct procedure for flambé work?

9. What are three safety points to consider when using (cooking on) a guéridon?

Chapter 13

Beverage equipment and service knowledge

Our guests come expecting a pleasant and satisfying dining experience. All too often the expectations built up by an exciting menu, all the good work of the chef in the preparation and presentation of the food and the good service of the food by the waiter are thrown away by a lack of professionalism in the preparation and service of the drinks.

On completion of this chapter you will have a basic understanding of the following:

❖ Beverage equipment identification

❖ Equipment preparation and maintenance

❖ Beverage lists

❖ The handling and placement of equipment

❖ The coordination of food and beverage service.

Beverages are as important as the food in the dining experience, so they should be given as much careful attention when they are being prepared and served as the food receives.

Beverage equipment identification

The service of beverages requires a wide range of equipment. The types of equipment used will vary depending on the tasks to be performed and the type of establishment. A waiter should be familiar with the full range of commonly used equipment, including all the equipment described in this chapter.

GLASSWARE

There is now an enormous range of commercial glassware available so that food outlets, from the simplest to the most traditional and expensive, can choose glassware to suit all their particular needs.

When selecting glassware, management will take various factors into account, such as size, shape, ease of handling and washing, durability and price. The glassware selected should, of course, be appropriate to the style of the establishment and its menu.

While the designs of the glassware available vary considerably from manufacturer to manufacturer, there are standard basic shapes which identify the glasses as belonging to the various classical types (see pages 124–125).

SERVICE EQUIPMENT

Many specialist devices and types of equipment have been produced over the years to help the waiter with the extraction of corks, the carrying of drinks and the cooling of beverages. However, the '**waiter's friend**' remains the recognised device used by waiters to extract corks. The design of the waiter's friend is available in many forms, but the fundamental components are as demonstrated below.

Use of the waiter's friend is explained in Chapter 15.

Preparation and maintenance of equipment

The exact procedures to be adopted for the service of beverages will depend on the type of establishment, the styles of service offered, and the availability of service station areas.

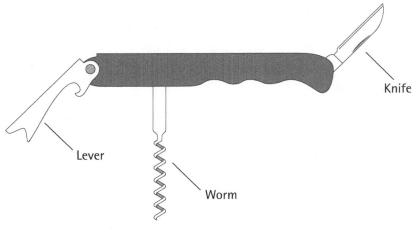

Knife

Lever

Worm

Waiter's friend (open)

Pre-service duties will include cleaning and polishing the necessary glassware, service station mise-en-place, preparation of ice buckets, and the handling and placing of equipment.

CLEANING AND POLISHING GLASSWARE

Although glasses are hygienically washed and sterilised by the high temperatures (not less than 77°C) of the washing cycle in a commercial dishwasher, it is still necessary to polish all glassware by hand before it is placed on the table or used to serve drinks. A lint-free polishing cloth should be used to polish glasses and make sure they are spotlessly clear.

SERVICE STATION MISE-EN-PLACE

Efficient service requires careful prior preparation of the service equipment. In some establishments this is done on a special piece of furniture in the dining room, known as the *drink waiter's station*. Often space doesn't allow this and a service station must be established behind the scenes; commonly, the bar is used.

Here is a checklist of supplies and equipment that may be required for beverage service:

❖ additional glassware

❖ drinks trays

❖ wine lists

❖ table napkins

❖ straws

❖ toothpicks

❖ service cloths

❖ docket book

❖ wine coolers

❖ ice buckets.

Paris

Red wine

White wine

Champagne flute

Champagne saucer

Balloon

Standard beer glass (200mL)

Pot/Middy (285mL)

Pint glass (470mL)

Standard glassware

Cocktail/Jockey (140mL)   Small cocktail (Martini) (90mL)   Colada/Pocco grande

Tumbler   Highball   Pilsner   Old Fashioned

Shot glass   Small liqueur   Port   Sherry

Standard glassware (continued)

WINE LIST

SPARKLING WINES & CHAMPAGNE

Australian	glass	bottle
Cockatoo Ridge Brut Reserve NV	$6	$25
1999 Croser Vintage Coonawarra		$60
1998 Domaine Chandon		$60
Fleur de Lys Pinot Chardonnay NV	$5	$20
Jansz Brut NV		$50

New Zealand	bottle
1997 Daniel Le Brun Blanc de Blanc	$80

French	
Louis Roederer Brut Premier NV	$120
Moet et Chandon	$120
Veuve Clicquot NV	$120

WHITE WINES

Australian	glass	bottle
2002 Chevalier Chardonnay	$5	$22
2002 Padthaway Chardonnay	$5	$22
2001 Houghton Gold Chenin Blanc	$7	$30
2002 Voyager Estate Chenin Blanc		$40
2002 Vasse Felix Classic Dry White		$35
2002 Yalumba Classic Dry White	$6	$25
2002 Tahbilk Marsanne	$6	$25
2001 Angas Busy Road Pinot Gris	$7	$30
2001 Leasingham Clare Riesling	$7	$30
2002 Lindemans Bin 75 Riesling	$5	$20
2002 Tahbilk Roussanne	$6	$25
2002 Lenswood Sauvignon Blanc		$45
2001 Lilydale Sauvignon Blanc		$40
2002 Peter Lehmann Semillon	$5	$20
1999 St Hallett Semillon		$35
2002 Tahbilk Verdelho	$6	$25
2002 Houghton White Burgundy	$5	$22
2002 Grants Gully White Burgundy	$5	$20

House White	glass	bottle
Chardonnay	$5	$20
Riesling	$5	$20
Sauvignon Blanc	$5	$20
White Burgundy	$5	$20

New Zealand	bottle
1998 Beach House Chardonnay	$30
2000 Fairhall Downs Chardonnay	$36
2001 Villa Maria Reserve Chardonnay	$60
1997 Bradshaw Estate Fumè Blanc	$40
1999 Felton Road Vin Gris (Pinot Gris)	$40
2001 Vavasour Riesling	$40
2002 Trinity Hill Roussanne	$50
2001 Cloudy Bay Sauvignon Blanc	$70
2001 Framingham Sauvignon Blanc	$50
2002 Giesen Sauvignon Blanc	$35
2002 Selaks Sauvignon Blanc/Semillon	$35

French	bottle
2001 Hugel Pinot Blanc	$35

RED WINES

Australian	glass	bottle
1998 Turramurra Estate Cabernet		$55
2001 Tahbilk Cabernet Merlot		$55
2000 Black Jack Cabernet Merlot		$50
2002 Rosemount Cabernet Merlot	$7	$30
2001 Punt Rd Cabernet Sauvignon	$8	$35
2001 Sandalford Cabernet Sauvignon		$50
2000 Vasse Felix Classic Dry Red	$8	$35
2003 Brown Brothers Dolcetto	$7	$30
2001 Tahbilk Malbec	$6	$25
2000 Yarra Ridge Merlot	$8	$35
2001 Grant Burge Hillcot Merlot		$40
2002 Ninth Island Pinot Noir		$45
2001 Stonier Pinot Noir		$45
2001 Maglieri Shiraz	$8	$35
2001 Saltram Classic Shiraz	$5	$20

House Red	glass	bottle
Cabernet Merlot	$5	$20
Cabernet Sauvignon	$5	$20
Merlot	$5	$20
Shiraz	$5	$20

New Zealand	bottle
1997 Seibel Wines Cabernet Merlot	$30
1995 Milness Cabernet Sauvignon	$30
1998 St Jerome Cabernet Sauvignon	$30
1998 Omaka Springs Merlot	$35
1998 Riverside Merlot	$35
2001 Schubert Pinot Noir	$90

French	bottle
1993 Chateau Duboeuf Merlot	$30
2000 Maison Guigal Cotes-du-Rhone	$35
2000 Abbotts Ammonite Cotes Rousillon	$40

DESSERT WINES

Australian	glass	bottle
2000 Brown Brothers Crouchen	$5	$20
2001 Tahbilk Dulcet	$5	$20
2000 De Bortoli Noble One (375mL)	$9	$45
1998 Woodbury Botrytis Semillon	$7	$30
1999 Yalumba Botrytis Viognier	$8	$40

French	bottle
1998 Chateau Laville Sauternes (375mL)	$60
2000 Muscat de Beaumes de Venise	$55

A wine list

BEVERAGES and WINES

- Beers
Cascade Premium	$8
Coopers Sparkling Ale	$8
Crown Lager	$8
Fosters Lightice	$8
Hahn Ice	$5
Toohey's Extra Dry	$8
Victoria Bitter	$5
XXXX Light Bitter	$5

- Coolers
Bacardi Breezer	$7
Lemon Ruski	$7
Vodka Cruiser	$7

- Aperitifs
Vermouths	$5
Campari	$5
Dubonnet	$5
Pernod	$8

- Spirits
Akropolis Oyzo	$8
Bacardi White	$8
Bundaberg Rum	$8
El Toro Tequila	$8
Gordons Dry Gin	$8
Smirnoff	$8
Jack Daniels	$8
Jim Beam	$8
Johnny Walker Red	$8
Jameson's Irish whiskey	$8

- Sherry
Penfolds dry sherry	$8
Harvey's Bristol Cream	$8
Tio Pepe	$8

- Brandys & Cognacs
Remy Martin	$8
Courvoisier VS	$8
Armagnac	$10

- Port & Fortified Wines
Campbell's Rutherglen Muscat	$8
Brown Brothers Very Old Port	$8
Penfold's Grandfather Port	$9
All Saints The Keep Tokay	$8

- Liqueurs
Bailey's Irish Cream	$8
Cointreau	$8
Dom Benedictine	$8
Drambuie	$8
Galliano	$8
Grand Marnier	$8
Kahlua	$8
Sambucco	$8
Tia Maria	$8

- Sparkling Wine & Champagne
	glass	bottle
Killawarra Brut	$8	$28
Sir James	$7.50	$24
Moet et Chandon		$150
Veuve Clicquot		$150

- White Wine
House white (dry)	$7	$25
Beach House Chardonnay	$8	$30
Ingoldby Chardonnay	$8	$30
Mitchell Watervale Riesling	$9	$35
Giesen Sauvignon Blanc	$8	$30
Lenswood Sauvignon Blanc	$9.50	$40
Moss Wood Semillon	$10.50	$60

- Red Wine
House red	$7	$25
Vasse Felix Classic Dry Red	$9	$35
Riverside Merlot	$8	$30
Rosemount Cabernet Merlot	$8	$30
Logan Cabernet Sauvignon	$9.50	$40
Milness Cabernet Sauvignon	$8	$30
Stonier Pinot Noir	$10.50	$60
Hardy's Shiraz Cabernet	$8	$30
Wynns Coonawarra Shiraz	$9.50	$45

- Dessert Wine
De Bortoli Noble One (375mL)	$10	$45
Woodbury Botrytis Semillon	$8	
Maglieri Lambrusco		$20

- Soft Drinks
Coke/Diet Coke/Lemonade/Sprite	$4
Juices—apple, orange, pineapple	$4
Lemon, lime & bitters	$5
Mineral water—flavoured	$4
Mineral water—sparkling/still	$4

A beverages and wine list

WINE COOLERS AND ICE BUCKETS

Ice buckets are used to keep white and sparkling wines cool in more formal and usually more expensive restaurants, while simple insulated wine coolers, sometimes placed on the table, are used in less formal establishments.

Ice buckets, when required for use, should be half filled with a mixture of crushed ice (two-thirds) and cold water (one-third). The water allows the bottle to sink into the ice instead of balancing on top of it. The bucket may be placed in a tripod stand.

Beverage lists

Beverage lists come in many different formats. Sometimes a so-called 'wine list' will contain the entire range of beverages available. This format can be confusing for the guest. A better solution is to divide the various different types of beverage into separate lists. This helps guests find and select the beverages they want more speedily. Possible lists may include:

❖ cocktail list

❖ drinks list (includes apéritifs, beers, spirits and non-alcoholic drinks)

❖ wine list

❖ after-dinner drinks list (liqueurs, ports, brandies)

❖ liqueur coffee list.

THE WINE LIST

Wine lists are usually divided into wines of different types—for example:

❖ sparkling wines

❖ white table wines

❖ red table wines

❖ dessert wines.

See pages 126–127 and Chapter 14.

Handling and placement of equipment

For reasons both of hygiene and presentation, it is essential that all glassware is handled by the stem or the base of the glass. When glasses are being moved in the presence of

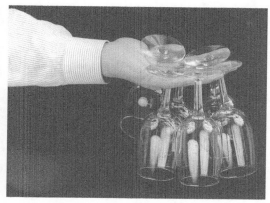

Carrying glasses in the presence of guests   Carrying glasses before the guests' arrival

guests, they should always be carried on a beverage tray. Before the guests' arrival, when the tables are being laid, several glasses may be held upside down in one hand with their stems between one's fingers.

PLACING OF GLASSES

If a single glass is being laid it should be placed 2.5cm above the main knife. If more than one glass is placed on the table, the glasses are positioned in a line at an angle of 45° in the order in which they will be required (see Chapter 4).

Food and beverage coordination

Regardless of whether the establishment has decided to employ a specialist wine waiter, or to make all its waiters responsible for both food and beverage service, it is essential that the service of the food and the beverages is coordinated.

The food waiter and the wine waiter must communicate if they are to provide a co-ordinated sequential service. The sequence of service requires both food and beverages to be served at the appropriate times throughout the meal without interfering with each other.

KEY POINTS IN FOOD AND BEVERAGE SERVICE COORDINATION

❖ Before the menu is presented, guests are offered an apéritif (pre-dinner drink) to stimulate the appetite.

❖ Because the wines are selected to complement the food chosen, the wine list is usually presented after the food order has been taken.

❖ The wine selected to accompany each course is served just prior to the food in that course. It is usual to serve white wines before red, dry wines before sweet, young wines before old—but what wines are chosen and in what order is, of course, up to the guest; the 'right' wine is what the guest wants.

❖ Remind the guests that dessert wines are available and offer the wine list prior to the dessert order being taken. Dessert wines are sweet, and complement sweet dishes.

❖ Orders for after-dinner alcoholic beverages are taken before coffee is served. This allows the coffee and other after-dinner drinks, such as port, cognac or liqueurs, to be served at the same time.

BYO restaurants

BYO restaurants—that is, establishments that allow guests to bring their own wine—are an accepted part of our industry, although they are more common in some states than in others. While, in BYO establishments, we don't take responsibility for the sale of the alcoholic beverages brought in by the guests, we are responsible for serving them as professionally as in a fully licensed restaurant.

QUESTIONS

1. What are the three main parts of a waiter's friend?

2. What factors are taken into account when purchasing glassware?

3. Draw each of the following: colada glass, pilsner glass and Old Fashioned glass.

4. What beverage mise-en-place items are required at the service station for beverage service?

5. What proportions of ice and water go in an ice bucket?

6. How are wines usually divided on a wine list?

7. How should glasses be handled?

8. Why is the wine list usually presented after the food order has been taken?

9. What is the right order in which to serve wines of different types?

Chapter 14

Beverage product knowledge

You may think that a bright personality and the ability to open a bottle of wine or beer is enough to make you a good drinks waiter, but without a basic knowledge of products, the job can only be half done.

Guests rely on us to assist them in making their selections and in educating them in their appreciation of wine and other beverages.

On completion of this chapter you will have a basic understanding of the following:

❖ Apéritifs

❖ Sparkling wines (including champagnes)

❖ Table wines

❖ Dessert wines

❖ Fortified wines

❖ Spirits

❖ Beers

❖ Liqueurs

❖ Cocktails

❖ Non-alcoholic beverages (except espresso coffee)

❖ Aerated waters.

You cannot be a competent drinks waiter without a basic knowledge of the products you are selling and serving. This chapter lists and defines the various categories of beverages.

To learn more about all the beverages served in Australian restaurants, we recommend that you consult *The Australian Bar Attendant's Handbook* by George Ellis. For a more detailed introduction to Australian wine, see *Australian Wine: Styles and Tastes* by Patrick Iland and Peter Gago; and for more information on cocktails, we recommend *The Australian Bartender's Guide to Cocktails* by Russell Steabben and Frank Corsar.

The styles and service of coffee and tea are described in Chapters 15 and 16. More information on particular beverage items may be found in the glossary.

Apéritifs

An **apéritif** is a pre-dinner drink taken to stimulate the appetite. Most apéritifs are dry in style because dry beverages stimulate the appetite, while sweet drinks tend to dull the appetite. In spite of this, some guests may prefer a sweet drink before a meal as an apéritif.

A good waiter will never make guests feel uncomfortable because of the drinks they have chosen, no matter how inappropriate they may seem. Popular apéritifs include:

❖ dry sparkling wine

❖ pre-dinner cocktails (acidic or dry, rather than creamy)

❖ dry sherry

❖ dry ('French') vermouth

❖ a proprietary apéritif (for example, Campari, Fernet Branca, Dubonnet or Rosso Antico).

Sparkling wines

Sparkling wines get their sparkle or effervescence from carbon dioxide. Carbon dioxide is produced naturally in the process of fermentation and can be retained to produce a sparkling wine.

Champagne is a region of France renowned for its sparkling wine, known as **champagne**. Champagne is made by a complex process called the méthode champenoise, or traditional method. The most important feature of the traditional method is that fermentation takes place in the bottle.

Some sparkling wines are made by the traditional method, while others ferment in vats and are bottled later.

Champagnes are made from a blend of grapes, usually pinot noir (black) and chardonnay (white) grapes. Meunier, a grape variety similar to pinot noir, is also used. Only sparkling wine produced in the Champagne district of France by the traditional method should be called 'champagne'. In Australia, sparkling wine was described as 'champagne' if it was fermented in the bottle. This is no longer done, but guests may still ask for 'champagne' when in fact they want an Australian sparkling wine.

If natural fermentation in a vat produces the sparkle a wine can be called a 'sparkling wine', but if the carbon dioxide is injected into the wine it becomes a 'carbonated wine'. Few carbonated wines are now made.

The styles of sparkling wine include:

❖ brut: dry

❖ sec: medium dry

❖ demi-sec: medium sweet

❖ doux: sweet.

The dry styles are much more popular than the sweet.

There are also red sparkling wines, usually described by the name of the grape variety used to make them—for example, 'sparkling shiraz'. Guests may refer to them as 'sparkling burgundy'.

Sparkling wines are sold in a variety of large bottle sizes, each with a special name. The most common are:

❖ magnum: two ordinary (750mL) bottles

❖ jeroboam: four ordinary bottles.

Other large bottle sizes are listed in the glossary (see 'champagne bottle sizes').

Table wines

The term 'wine' indicates a type of beverage made from fermented fruit. Wine may be made from a variety of fruits, but wine as we generally know it is made from the fermented juice of grapes. When another fruit is used to produce the wine, the name of the fruit used is included on the label (for example, 'strawberry wine').

As a general rule, red wine is made from 'black' (actually, purple) grapes, while white wine is made from 'white' (actually, green) grapes. Rosé wine, which is pink (rosy), is made from black grapes, but the skins of the grapes are removed early in the process of fermentation.

In the early days of the Australian wine industry, most wines were given generic labels—that is, they were described as wines of a particular type, such as moselle, hock, chablis, burgundy or claret. Many of these are the names of districts in France or Germany famous for producing a particular kind of wine. These traditional generic names are no longer used on Australian wine labels.

Nowadays, bottled wines usually carry a varietal label—that is, the label tells you the variety of grape from which the wine is made. Common white varietal wines include chardonnay, riesling, sauvignon blanc, sémillon and gewürztraminer. Cabernet sauvignon, pinot noir, merlot and shiraz are red varietal wines. Often two or more varieties are blended: for instance, traminer-riesling or cabernet-merlot.

The different varietal wines have distinct characters. For example, chardonnay is a fruity, dry, strongly flavoured wine; riesling, also (usually) a dry wine, has a lighter taste; sauvignon blanc is lighter still and more acidic; and gewürztraminer is distinctly spicy. Similarly, among the reds, cabernet sauvignon wines are dark red and have a strong flavour with an astringent edge to it; merlot wines, though dark red, are much softer, and pinot noir wines are light and velvety.

In addition to the varietal description, wine labels often state the district the grapes used to make the wine came from—for example, Coonawarra, Hunter Valley, Margaret River, Barossa Valley, Goulburn Valley, Yarra Valley.

As a general rule, red wines should be served at 'room temperature' (about 18°C) and white wines should be served mildly chilled

Table wines

(about 6°C). Some old table wines that have developed a crust or sediment in the bottle may need to be decanted before they are served.

Dessert wines

Dessert wines are rich and sweet. They are designed to be consumed with sweet food items. Many sweet dessert white wines are made from grapes that have been affected by a kind of mould called botrytis, or noble rot. Sauternes is a famous French white dessert wine made from botrytis-affected grapes. De Bortoli's Noble One is a well-known Australian example. Spätleses and ausleses are German styles of dessert wine.

Fortified wines

A fortified wine is a wine strengthened with the addition of spirit. The spirit also preserves the wine for longer periods after the bottle is opened. Fortified wines include sherry, port, vermouth and muscat.

There are three main styles of sherry: fino, amontillado and oloroso. Finos are light and dry, amontillados are darker and richer, while olorosos are full, rich and sweet. The sherries taken with a meal are usually dry. They are used as apéritifs or taken with soup.

Port comes in several styles: ruby, tawny and vintage. Ruby ports are the youngest and least rich. Vintage ports mature in the bottle and develop a sediment. They should be decanted before they are served. Almost all ports are red wines, but there are white ports.

The terms 'sherry' and 'port' are still used in a generic sense in Australia to describe wines of a particular type. However, sherry is really a Spanish fortified wine from a district called Jerez, and real port comes from Oporto in Portugal. Australian winemakers are now dropping the generic descriptions from their labels and using instead a description of the style—'amontillado' or 'tawny', for example.

Vermouth comes in three main styles: rosso, bianco and dry. Rosso vermouth, sometimes called 'Italian', is light red, sweet and strong-tasting. Bianco is light golden, medium sweet and spicy. Dry or 'French' vermouth is almost colourless and has a delicate flavour.

Australian muscats are rich, sweet, fortified wines. They can be red ('brown') or white ('orange'). Because they are often made from a variety of grape called *muscat de Frontignan*, these wines are sometimes called frontignacs. Rutherglen in Victoria is famous for its muscats.

Spirits

Spirits are distilled alcoholic beverages. Distillation is the process of converting liquid into vapour by heating, and then condensing the vapour back to liquid form. Almost any fruit or vegetable can be crushed to liquid, fermented and then distilled to make a spirit.

These are the most popular spirits and their base ingredients:

Spirit	Base
❖ whisky	grain (barley, wheat or maize)
❖ gin	neutral spirit made from grain and then flavoured with juniper berries
❖ rum	sugar cane
❖ vodka	potatoes or grain
❖ brandy	grapes
❖ tequila	blue Mezcal (a Mexican cactus)

Scotch has long been the most popular kind of whisky but its popularity is now almost equalled by bourbon. Bourbon is an American whiskey made from maize. Cognac and armagnac are high-quality French brandies.

Beer

Beer is made from fermented grain by a process called *brewing*. The traditional ingredients are malt (barley soaked to germinate and then dried), yeast, hops and water. Beer is the general term for ales, lagers and stout. Ales and lagers are made by different techniques of fermentation: ales are top-fermented, whereas lagers are bottom-fermented. In general, lagers are paler and more highly carbonated than ales. Most Australian beers are lagers. Stout is a dark, heavy beer. Guinness is a kind of stout. Draught beer is beer drawn from a keg, rather than bottled or canned.

Australian beers often carry confusing labels: for example, a lager can be described as 'ale', and a bottled or canned beer may be described as 'genuine draught'!

Liqueurs

Liqueurs are spirit-based (or sometimes wine-based) liquors, sweetened and flavoured. They have been made for centuries and new ones are being devised all the time.

Some liqueurs are generic; that is, they are liqueurs of a particular type, which anyone may make. Advocaat, crème de menthe and curaçao are popular generic liqueurs. Other liqueurs are proprietary; that is, they may only be made by a single distiller who owns the right to make the liqueur of that name. Bénédictine, Cointreau, Drambuie and Grand Marnier are examples of proprietary liqueurs.

Liqueurs are often taken with the coffee at the end of a meal. They are usually served neat (without any mixer) in a liqueur glass or on ice in an Old Fashioned or cocktail glass. They may also be taken in black coffee, as a liqueur coffee (see page 138 and Chapter 15).

Liqueurs are also frequently used in cocktails.

Cocktails

Cocktails are mixed drinks. Two or more ingredients are mixed by one of the following methods:

❖ shake and strain (in a cocktail shaker, with ice)

❖ stir and strain (in a mixing glass, with ice)

❖ blend (in an electric blender, with the quantity of ice specified in the recipe)

❖ build (prepared directly in the glass).

Cocktails fall into three broad types:

❖ *Pre-dinner cocktails:* These are usually acidic or dry and make good apéritifs. A Dry Martini is a classic pre-dinner cocktail.

❖ *After-dinner cocktails:* These tend to be richer, often creamy and sweet. A Brandy Alexander is an example.

❖ *Long-drink cocktails:* These often contain fruit juices, soft drinks or milk, in addition to their alcoholic base. A Tom Collins is an example.

Most cocktails are served with a garnish, which may be as simple as a slice of lemon or a cherry. The garnish should be appropriate to the contents of the cocktail. Many classic cocktails are served with a standard garnish. For example, a Dry Martini is served with an olive or a twist of lemon; whereas a Gibson, which otherwise has the same ingredients, is served with a pearl onion. The garnish is as much part of the cocktail as its liquid ingredients.

Most classic cocktails have a spirit base, but this isn't necessary. Some cocktails contain no alcoholic component at all, in which case they are called virgin cocktails, or mocktails (see below).

Non–alcoholic drinks

The term 'non-alcoholic' drinks includes a wide variety of beverage items, from cold to hot and from the simple to the exotic. Some are served from the kitchen/still area and some are dispensed from the bar.

NON-ALCOHOLIC DRINKS SERVED FROM THE KITCHEN/STILL AREA

Tea

A growing demand for a variety of specialist teas—fruit, herbal and decaffeinated styles—has made it necessary for establishments to expand the range of teas they offer.

The majority of teas are prepared using teabags, but some teas are prepared from loose leaves, sometimes using a tea plunger. When tea is served, guests pour for themselves. The waiting staff provide the prepared tea, additional hot water, milk, lemon and appropriate sweeteners. Types of tea commonly requested include English Breakfast, Earl Grey, Darjeeling, China and herbal tea.

Iced tea is increasingly popular. It consists of tea poured over ice and served cold in a glass, garnished with slices of lemon or mint leaves.

Coffee

Coffee is made from the seeds of the *Coffea* tree. The seeds are roasted and then ground to a powder, which is used to make the familiar drink we all know.

Because of its special importance in modern hospitality establishments, espresso coffee is dealt with in a separate chapter of its own (Chapter 16).

Not all coffee is espresso coffee. There are other ways of making coffee apart from the espresso machine—for example, by the use of filters or plungers. And, of course, there are instant coffees.

There are various different coffee blends, and the tastes of coffee made with beans from different countries (Kenya, Brazil, Papua New Guinea, etc.) can be detected by experts, but it is unlikely that the ordinary restaurant waiter will be asked for a particular blend of coffee. Apart from ordinary coffee, there are other types of coffee or ways of serving them for which you may be asked.

❖ **Decaffeinated coffee** is real coffee with the stimulant caffeine extracted. Some people prefer coffee substitutes, such as Caro, which are not made from coffee beans.

❖ **Liqueur coffees** are those served with a spirit or liqueur. Irish coffee is the obvious example. Irish whiskey is added to hot, sugared black coffee and then topped with cold, fresh cream. A number of different liqueurs, such as Tia Maria and Galliano, are commonly served with coffee in the same way, under a variety of names—for example, 'Jamaican coffee' or 'Roman coffee'. (Liqueur coffees are, of course, alcoholic, but for convenience are described here in the coffee section. See also Chapter 15.)

❖ **Iced coffee** is strong, black coffee mixed with cold milk and usually garnished with ice cream.

Where coffee isn't made with an espresso machine, you may be asked for 'black' coffee (with no milk) or 'white' coffee (with milk). White coffee is sometimes described by its French name '*café au lait*' (coffee with milk).

The service of coffee is discussed in Chapter 15 and also in the chapter on espresso coffee (Chapter 16).

Hot chocolate

Nowadays, chocolate is almost always made from pre-prepared (instant) powder. The quality of the chocolate used is the essential factor in the end product—the better the brand, the better the result. If good chocolate is to be served, you must use a good product—Suchard, for example. The chocolate powder is mixed with hot (but not boiled) milk before service, usually in the cups in which it is to be served (see Chapter 15).

NON-ALCOHOLIC DRINKS DISPENSED FROM THE BAR

Aerated waters

Aerated water is simply water charged with gas, usually carbon dioxide, to make it effervescent. It often contains a syrup for taste and colour. Soda water is a colourless and tasteless aerated water. Bitter lemon, Indian tonic and dry ginger are flavoured aerated waters.

Fruit juices

Commercially packaged brands (canned or bottled) may be used or, for some varieties of fruit such as oranges, lemons and grapefruit, the juices may be prepared fresh. Serve chilled. Ask your guests whether they prefer their fruit juice with or without ice.

Squashes

These are preparations of fruit juices or syrups with sugar, water and other ingredients, usually described by the manufacturers as 'cordials'. Some use mineral water instead of ordinary tap water.

Mineral waters

Mineral waters are so-called because of the minerals they contain, which are said to be good for health. They may be still or sparkling (effervescent). If they come unadulterated from a natural spring, they are 'natural' mineral waters; if their effervescence is also natural, rather than added by the injection of carbon dioxide, they are 'naturally sparkling'. Many mineral waters are not 'natural' but are simply tap water with minerals in it or added by the manufacturer. Aerated mineral waters are made from purified waters with carbon dioxide added.

Natural mineral waters are often named after their place of origin. Evian, Perrier and Vichy are natural mineral waters from France. (Evian is still, while Perrier and Vichy are naturally sparkling.) Appollinaris is a naturally sparkling variety from Germany. There are several Australian mineral waters, such as Hepburn Spa and Deep Spring. Bisleri is an Italian brand, made under licence in Australia. Sometimes mineral waters are flavoured with fruit juices.

Mineral waters can be used as mixers or appreciated as refreshing, cleansing beverages.

Non-alcoholic wines

These are prepared from a fruit juice base and can be still or aerated. They contain no alcohol.

Non-alcoholic cocktails (virgin cocktails or mocktails)

Current trends towards responsible drink-driving habits have led to an increase in the availability of non-alcoholic drinks, including many **virgin cocktails** and other non-alcoholic mixed drinks. Often these are simply 'virgin' versions of an already popular cocktail—that is, the alcoholic ingredient is omitted or some other ingredient is substituted for it. A Virgin Mary, for example, is a non-alcoholic variation of a Bloody Mary, with the vodka omitted. Responsible beverage servers are now very conscious of the need to offer a good variety of non-alcoholic drinks.

QUESTIONS

1. What is an apéritif?

2. What is a varietal wine?

3. At what temperature should red wine be served?

4. At what temperature should white wine be served?

5. What is a fortified wine?

6. What style of fortified wine is an amontillado?

7. What makes the bubbles in sparkling wine?

8. What is the difference between wines and spirits?

9. What is the difference between lager and stout?

10. Describe the following methods for making cocktails: shake and strain, blend, and build.

11. How do you make an iced coffee?

12. What is a 'naturally sparkling' mineral water?

Chapter 15

Beverage service procedures

Do patrons purchase drinks simply to quench their thirst?
Or do they really come to enjoy a total hospitality
experience—the company, the atmosphere of the venue,
and the service—as well as the food they eat and the
beverages they drink?

On completion of this chapter you will have a basic understanding of the following:

❖ Responsible service of alcohol

❖ Selling beverages

❖ Taking beverage orders

❖ Handling glassware

❖ Tray carrying and service

❖ Beer service

❖ Wine service

❖ Sparkling wine service

❖ Liqueur, port and brandy service

❖ Changing glassware

❖ Tea and coffee service.

Beverages may be served as an individual item of service (for example, in bar or lounge service), or their service may be carefully coordinated with the service of the food so that the beverages complement the food, enhancing the guests' enjoyment of both.

The style of beverage service offered will depend on the character of the establishment and the type of beverages being served. The methods of service explained in this chapter reflect a commonsense approach to beverage service, but some establishments may require variations on these procedures.

Responsible service of alcohol

Everybody in the industry, both management and staff, is expected to have a responsible, caring and professional approach to the serving of alcohol on licensed premises. All employees of licensed premises should be trained in the responsible service of alcohol (RSA) to ensure that a safe and friendly environment is provided for all guests and staff, and that the venue abides by its legal obligations in accordance with the state liquor Act.

The principles of RSA are as follows:

❖ Alcohol should not be served to minors.

❖ Drunken or disorderly people should not be allowed on the premises.

❖ Alcohol should not be served to anyone who is intoxicated.

❖ There should be no promotions that encourage binge drinking or drunkenness.

The signs of intoxication are:

❖ a noticeable change in behaviour

❖ slurring or mistakes in speech

❖ clumsiness, knocking things over or fumbling with change

❖ loss of coordination

❖ confusion, delays in responding

❖ the smell of alcohol on the breath.

Clumsiness is often thought of as a transition stage on the way to serious intoxication.

A number of strategies can help prevent customers from becoming intoxicated. For examaple:

❖ Train staff in the responsible service of alcohol.

❖ Avoid unacceptable alcoholic drink promotions.

❖ Promote low-alcohol and non-alcoholic products.

❖ Suggest food to people who seem likely to have too much to drink. (Having something to eat should slow down their drinking.)

❖ In a function environment, ask guests before topping up their alcoholic beverages.

❖ Make water and non-alcoholic beverages readily available at all times.

If people do appear to be becoming intoxicated, suggest transport options—for example, offer to call them a taxi, or ask one of their friends to take them home.

UNDERAGE PATRONS

The details of the licensing laws differ slightly from state to state, and the state liquor Acts are altered from time to time. It is essential that anyone working on licensed premises have an understanding of the law as it applies where they are working. As a general rule, minors (young people under the age of 18) must not be supplied with alcohol, nor are they allowed to remain on licensed premises unless they are accompanied by a parent or guardian, or they are there to eat in a restaurant or for some other necessary reason, such as for training or they are employed on the premises in some role that doesn't involve them in handling alcohol.

If you are in doubt about a person's age, you must ask for reliable evidence that they are not underage before serving them alcohol or allowing them to consume alcohol on the premises. Acceptable evidence of age can include any of the following:

❖ driver's licence with photo

❖ passport

❖ proof of age card

❖ keypass card.

PENALTIES

There are severe penalties for both management and staff for those who:

❖ supply liquor to an intoxicated person

❖ permit drunken or disorderly people to enter or to remain on licensed premises, or

❖ supply alcohol to people under the age of 18.

The penalties vary from state to state. You must check with your State Liquor Licensing Commission for further information.

Selling beverages

The way beverages are sold is a key part of the hospitality experience. Bar and beverage service staff must not only understand the products they are selling, but must also know how to serve them, from both technical and psychological points of view.

VENUES OFFERING BEVERAGE SERVICE

Venues offering beverage service include:

❖ bars (both public and specialised)

❖ lounges

❖ restaurants/cafés

❖ function facilities.

Bars

The traditional names for the various bars suggest different types of customers, products and skills in the bar staff. For example, the traditional public bar was a male domain and most people drank beer there. Public bar staff needed few technical skills beyond knowing how to pull a beer and how to mix basic spirits. Things have changed. Today's public bar attracts both sexes; it may offer entertainment and a complete range of beverages including premium beers, cocktails and wines. Staff in public bars, saloon bars, wine bars, cocktail bars and licensed coffee bars must now be equipped with the skills of mixing cocktails and serving wines, and must know how to sell them.

Techniques and vehicles for selling beverages in bars include verbal recommendations, promotional displays, board listings and the availability of a drinks list.

Lounges

Lounge services are offered in a wide diversity of venues, including pubs, hotels, restaurants, lounge bars, wine bars, discos and clubs. Guests can sit away from the bar in a relaxed environment and may be served by floor staff.

In the lounge, sales are made by verbal recommendation or by using sales tools such as tent cards or cocktail/beverage lists. In lounges, the 'club service' procedure is normally used (see page 148).

Restaurants/Cafés

When beverages are sold and served in a restaurant/café, consideration has to be given to the integration of food and beverage service so that they enhance each other.

The methods used to sell beverages in restaurants/cafés vary to suit the type of establishment, but they commonly include verbal recommendation, tent cards, cocktail/beverage lists, wine lists and liqueur coffee lists.

Functions

The beverages served at a function are usually prearranged by the function host with the venue operator, rather than sold at the function itself. The venue manager may make verbal recommendations or may offer a beverage list from which the host selects the drinks to be served. Should the host request a beverage not on the standard list, it may be purchased by the venue especially for the function.

GENERAL POINTS ON SELLING BEVERAGES

When selling beverages:

❖ Don't dictate your personal preferences.

❖ Offer a diversity of recommendations so that guests are prompted to choose what they personally prefer.

❖ Suggest beverages that complement the occasion, but don't convey any sense of disapproval if something 'unsuitable' is chosen. Guests have the right to drink whatever they choose, and have come to enjoy themselves, not to be 'corrected'. Your job is to make them feel comfortable and relaxed.

Taking beverage orders

Beverage orders should be taken as soon as guests are comfortably settled, be it at a bar, in a lounge, or at a table in a restaurant/café. Remember the following points:

❖ When taking the orders (verbally or in writing), make sure you clearly understand them and that they are precise, so that the guests receive what they have ordered. If you are not quite clear what has been ordered, don't hesitate to confirm it with the guest.

❖ If there are several guests, write down the orders in logical order (as the guests are seated, or with some other clear identification) so that you place the drinks correctly when they are served (see Chapter 6).

❖ Avoid the use of abbreviations; they can easily cause confusion.

❖ Different venues have different methods for recording sales. Whatever the system in use—handwritten dockets, a cash register or a computer system—it is essential that you record all items sold in the appropriate way.

❖ In a restaurant, the wine order is usually taken after the guests have selected their food. Additional orders for wine may be taken throughout the meal; indeed, if the guests' glasses look nearly empty, discreetly ask the host whether another bottle of wine should be brought.

❖ The order for after-dinner beverage items (such as port, brandy or liqueurs) should be taken prior to the service of coffee, so that the drinks can be served with the coffee.

Glassware and drinks trays

All glassware, whether clean or used, should be carried upright on a drinks tray, except that wine glasses may be carried upside down by their stems, held between the fingers of the left hand, when setting covers before guests arrive (see Chapter 13).

All glassware (clean or used) must always be held only by the stem or base of the glass, never by the rim.

TRAY SERVICE

A drinks tray is carried on a slightly cupped hand on the pads of the fingers. This allows the waiter to adjust the balance of the tray as glasses are removed or added to the tray. When serving drinks to seated guests from a tray, hold the tray behind the guest's head and use your right hand to place the drink from the guest's right (see also Chapters 5 and 13).

Carrying the drinks tray

CLUB SERVICE

This form of service is usually provided only in superior cocktail bars and lounges. It is a high-quality service requiring special attention to detail. Napkins are placed on the table (or sometimes the bar) in front of each guest with their logos facing the guest. These napkins take the place of coasters and the glasses are placed on them. Mixed drinks are served with the spirit already poured into the glass. Mixers are served in small bottles or splits. The mixers are placed to the right of the glass so that guests can themselves add exactly the amount of mixer they wish. Beers and wines are poured by service staff from the bottle. In club service, nothing is on tap. Nuts or nibbles are normally placed on the table or bar and are included as part of the service without additional charge.

Tray service

Beer service

Beers, in particular the increasingly popular premium and imported beers, are often served at the table from individual small bottles/stubbies.

Imported or boutique beers may be served in a specific glass badged with the brewery's name.

BOTTLED BEER SERVICE

The procedure for serving individual bottles of beer at table is:

❖ Carry a clean cold beer glass and an open bottle to the table on a service tray.

❖ Hold the tray behind the head of the guest to be served.

Beer glasses

Pilsner glass

Standard beer glass

❖ Pick up the glass at its base and place it to the guest's right.

❖ Take the bottle in your right hand with the label clearly showing to the guest.

❖ Pour the beer into the glass on the table so that the flow is directed to the inside opposite edge of the glass. Pour slowly so that a head can form.

❖ Continue pouring until the glass is full, with a well-rounded head.

❖ If the bottle still contains some beer, place it on the table to the right of the glass, with the label facing the guest.

Sparkling wine service

The service of wine, whether it be over a bar or at a table, is a sequence of tasks involving basic skills. At a table, in licensed restaurants of all types, the correct procedure doesn't vary. There may be some modification of the procedure, however, in BYO restaurants and in busy bars.

❖ If a sparkling wine is selected, place suitable glasses (narrow 'champagne flutes') in position.

❖ Present the bottle on a service cloth held on the flat of your left hand, with the label directed to the host so that it can easily be read. Identify the wine verbally so that the host can confirm that the correct wine has been brought—'Your Domain Chandon, Sir?'

❖ When the host has confirmed that the wine is the correct one, proceed to open the bottle. Extraordinary pressure can build up in a bottle of sparkling wine, especially if it is shaken. People can be injured by the cork exploding violently out of the bottle, so opening must be undertaken with

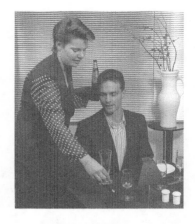

Serving beer at the table

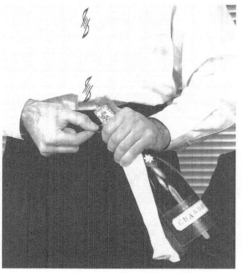

care. Never allow the bottle to point at your face or at anyone else; make sure it is pointing at the ceiling. During the opening procedure, hold the bottle at an angle rather than straight up. This reduces the pressure on the cork.

❖ Take the bottle firmly in your left hand, holding it at a 45° angle at waist height. With your right hand locate the wire ring on the **muselet** (muzzle or cage) and untwist it.

❖ Remove foil and the cage, holding the cork in place with the thumb of your left hand as an extra precaution.

The bottle may be held with the thumb in the punt

The bottle may be collared with a napkin

❖ Take a service cloth in the palm of the right hand and with it cover and firmly hold the cork. Hold the base of the bottle with your left hand and twist it to loosen the cork. Ease the cork gently out of the bottle into the palm of your right hand.

❖ Wipe the lip of the bottle with the service cloth.

❖ Hold the bottle in the right hand so that the label faces the host, and pour about 30mL of the wine into the glass for approval.

❖ Alternatively, the bottle may be held with the thumb in the punt (the indentation at the base) with the fingers spread out to support the body of the bottle.

❖ When the host has approved the wine, fill the guests' glasses (no more than two-thirds full), starting with the guest next to the host and moving around the table. Complete the service by topping up the host's glass. You may have to pause while pouring to allow the bubbles to settle.

❖ Place the bottle in an ice bucket unless the host requests otherwise. The bottle may be collared with a napkin to enhance its presentation.

❖ Top up the guests' glasses. When the bottle is empty, ask the host whether they require another bottle of the same wine, or whether you should bring the wine list for them to select another.

Table wine service procedures

An increasing number of wines now have a screw top (**stelvin screwcap**) replacing the traditional cork and seal. Except that the technique of opening the bottle is different, the procedure for serving a still table wine is essentially the same as that for a sparkling wine.

❖ Present the wine that the host has selected on a service cloth held on the flat of your left hand, with the label directed to the host so that it can easily be read. Identify the wine verbally, mentioning the company and variety, so that the host can confirm that the correct wine has been brought—'Hankin Cabernet Shiraz Malbec, Sir?' Don't open the bottle until the host has confirmed that the wine is the correct one. If the wine has been selected by a guest on behalf of the host, in that circumstance the waiter would direct the opening procedures to that person.

❖ When the host (or the host's guest) has confirmed the wine is correct, take the bottle firmly in your left hand, holding it at a 45° angle at waist height. Assuming that the bottle has a cork, and not a screw top, cut the foil with the blade of the waiter's friend just above the raised ridge about 5mm below the top of the bottle. Ease off the top of the foil with the point of the blade. Note that the foil should always be cut in this way even if a perforated pull-tab has been built into the foil; a clean cut prevents the wine from catching in the perforations and dripping when poured.

❖ Close the blade of the waiter's friend and open the spiral (corkscrew). Hold the neck of the bottle firmly in the left hand. Insert the sharp tip of the spiral into the centre of the cork. Slowly turn the spiral in a clockwise direction, keeping it in line with the core of the cork. Stop turning the spiral when the last turn of the spiral is still visible to prevent the spiral from piercing the base of the cork.

❖ Tilt the arm of the waiter's friend so that the lever rests on the lip of the bottle. Hold the lever in place using the side of your index finger. Now extract the cork by raising the opposite end of the body of the waiter's friend, exerting leverage on the lip of the bottle until the cork starts to bend.

❖ Now cease the lifting action and place your thumb and index finger at the base of the cork. Twist the cork gently onto its side to remove it from the bottle. This technique will prevent the cork from breaking and will allow it to be extracted without any distracting popping sound.

❖ Remove the cork from the spiral and return the waiter's friend to the pocket.

❖ In some establishments the cork is then presented to the host for inspection, particularly when fine red wines are being served. If it is the custom of the establishment to present the cork, a suitable small plate should be placed on the table in advance to accommodate it. If the cork has the name of the winery printed on it, the cork should be placed so that the writing can easily be read by the host.

❖ When opening a bottle with a screw top (stelvin screwcap), the removal is performed by holding the bottle firmly at a 45º angle at waist height, and turning the base of the

bottle in a clockwise direction to break the seal. The screwcap is removed and the procedure for service for bottles with a screw top as with any other bottle is as follows.

❖ Wipe the lip of the bottle with a service cloth to remove any dust particles.

❖ The bottle is then held firmly in the right hand with the label directed towards the host. Pour about 30mL of wine into the host's glass for approval. The wine should be poured into the centre of the glass with the bottle held above and not touching the glass. (Note: If the table is laid with more than one style of wine glass, the smaller glasses should be used for white wine and the larger for red.)

❖ After the host has approved the wine, fill the guests' glasses (no more than two-thirds full), starting with the guest immediately next to the host. At the table the wine is poured from the guests' right. Moving around the table, complete the service by topping up the host's glass.

❖ When all the glasses have been filled, bottles of white wine are placed in a cooler or ice bucket (if available), or placed on the table if the guests so request. Bottles of red wine should be placed on the waiter's station (if there is one) or placed on the table (if there is not, or if the guests request it).

❖ Bottles of red wine may be collared with a napkin to improve their presentation. White wine bottles, if placed in an ice bucket, may have a napkin draped over them.

❖ The discarded cap, foil and cork must not be left at the table or dropped into the ice bucket. Put them in your pocket and dispose of them at the bar.

❖ Keep an eye on the guests' glasses. When they are only one-third full, request if the guest would like the wine topped up.

❖ When the bottle is empty, ask the host whether they require another bottle of the same wine, or whether you should bring the wine list for them to select another. If another bottle of the same type of wine is ordered it is not necessary to change the glasses unless you are asked by the host to do so. Nor, in most (but not all) establishments, is it necessary to repeat the tasting procedure. Simply open the bottle and continue to top up the guests' glasses as before. If a different wine is selected, the glasses must be changed (see below) and the tasting and opening procedures are repeated in the same way as for the original bottle. The tasting procedure should also be repeated if the wine originally selected was an expensive one.

Bottles of red wine may be collared to improve their presentation

Red wine glass   White wine glass

Liqueurs, port and brandy

Liqueurs, port and fine brandies (usually cognac, but sometimes armagnac or old Australian brandy) are commonly served with coffee at the conclusion of a meal. The waiter may suggest them verbally or offer the drinks list after the sweets or desserts. Glasses are usually pre-poured and served at the table off a tray, but in some establishments the bottle is presented at the table and then poured for the guest or served from a beverage trolley.

| Brandy balloon | Liqueur | Small liqueur | Port |

The normal measure for brandy or liqueurs is 30mL, and for port or other fortified wines it is 60mL.

Fine brandies are served in wide-bottomed glasses (balloons). It is usual for guests to warm their own brandy glasses in the palms of their hands, but the waiter may be requested to warm the glasses before the brandy is served. Some establishments have special warming lamps for this purpose. Note that the glass is empty when warmed on a lamp, the brandy itself is not warmed for the guest.

Changing glassware

The original glassware may be used if additional bottles of the same type of wine as was originally ordered are required, unless a change of glasses is requested. If wine of a different style or type is ordered, then fresh glasses must be placed before the new wine is served.

Select the style of glassware appropriate to the style of wine chosen (for example, champagne flutes for a sparkling wine) and take the glasses to the table on a drinks tray. Place the glasses a little further away from the guests than the original glasses and at an angle of 45° to the right of them.

Remove the original glasses when the guests have finished the wine in them. Remember that the glasses should be handled by their stems whether they are being placed or removed.

After-dinner drinks (liqueurs, ports and after-dinner cocktails) should be served in fresh glasses for each new order.

Tea and coffee

The exact procedure for serving tea and coffee at the table will vary, depending on the venue, the style of service and the equipment available.

Tea and coffee service

The preparatory steps for tea and coffee service are:

❖ Take the order.

❖ Place the accompanying items (milk, sugar—white, coffee crystals or sweetner—and, if required, cream and lemon) on the table. An underliner may be used to present these items. Make sure the sugar bowl has a clean spoon with it.

❖ Place a cup and saucer, and a teaspoon, from each guest's right. If coffee or tea is being served with cheese or dessert, or for afternoon tea, the cup and saucer are placed to the right of the cover.

❖ If tea or coffee is being served at the end of a meal after the table has been cleared, the cup and saucer should be placed near the centre of the cover.

❖ The handle of the cup should be to its right, and the teaspoon should be placed at a 45° angle on the saucer just behind the handle.

TEA

The ritual for serving tea allows the guests to pour for themselves. Present the teapot and its accompanying hot-water pot on an underplate with a tea strainer and a small napkin.

The strainer and the napkin may not be provided in all establishments. A strainer is unnecessary if tea bags have been used, but one still may be asked for. If tea bags have been used, they should not be 'jigglers'; pots should not be served with tea-bag strings and labels hanging out. The napkin is to assist the guest to hold the hot pots while pouring.

Place the underplate to the right of the guest, with the handles of the teapot and hot-water pot directed to the guest.

COFFEE

The way in which coffee is prepared and served depends on which kind of coffee it is. Coffee was briefly described in the previous chapter, and Chapter 16 deals in detail with the making and service of espresso coffee.

Regardless of which kind of coffee is being served, there are no points to be won by serving insipid, thin coffee. Most people prefer rich and flavoursome coffee, with only a minority preferring their coffee weak. Most people who drink black coffee, particularly short black espresso, want quality, rich and strong coffee. Coffee can always be diluted if it is too strong. However, unless it is made with a plunger, it cannot be strengthened if it has been made too weak. The amount of water pushed through the coffee determines its strength, without (up to a point) compromising its richness and aroma. The less water, the richer and stronger the coffee.

Serving coffee from a coffee pot with a long spout

Service of coffee

Except when guests have ordered a full pot of coffee, or when espresso coffee is being served, coffee is normally poured for them at table.

Coffee is served from the right of the guest. If the coffee pot has a short spout (for example, a Cona pot), pick up the cup and bring it up to the pot to fill it. If the pot has a long spout, enabling you to direct the flow accurately, you should pour the coffee straight into the cup on the table.

In formal silver service, present the coffee pot and milk jug on an underliner and then pour both coffee and milk for the guests.

Coffee and tea can both be prepared on and served from a guéridon. If speciality coffees and teas are served, the guéridon enables you to make the most of their presentation (see Chapters 10 and 12). Liqueur coffees are particularly suited to preparation on the guéridon.

Hot chocolate

Hot chocolate is usually served in cups prepared in the still area, rather than poured from a pot at the table. The cups are served from the guests' right. For chocolate preparation, see Chapter 14.

Liqueur coffee

Liqueur coffees can provide an additional highlight to the dining experience. When prepared and served correctly, they will not only impress the guests but encourage them to order more. The simplicity of the technique involved, and its impressive visual effect, make liqueur coffees particularly appropriate for preparation in front of guests, either on a guéridon or at the bar.

The preparation of a perfect liqueur coffee is a simple process in which fresh, cold, pouring cream is floated on hot liqueured coffee. The result should be a dramatic visual contrast of white on black.

Pouring from a spirit warmer

A liqueur coffee is not only enjoyed for its eye appeal, aroma and taste; there is also the special sensation of drinking the hot liqueured coffee through the layer of cold, fresh cream. If a liqueur coffee is to be fully appreciated, it is essential that the cream should not be semi-whipped, whipped or dispensed from a pressure-pack can.

Although different establishments may make minor variations to the recipes, and may use their own names for the various liqueur coffees they offer, the simple technique used to create the perfect liqueur coffee shouldn't be altered.

LIQUEUR COFFEE PREPARATION

To prepare a perfect liqueur coffee:

❖ Choose an appropriate glass, preferably one made of clear, toughened glass, with a stem or handle.

❖ Pour 30mL of the selected spirit or liqueur into the glass.

Adding aerated pouring cream

❖ For additional visual effect, the spirit or liqueur can be warmed and flamed in a spirit warmer before it is poured into the glass.

❖ Add sugar, if required.

❖ Pour hot, black, percolated or filtered coffee into the glass, filling the glass up to 1.5cm from its top.

❖ Stir.

❖ Lightly aerate the cream by shaking it in a closed container for about three seconds.

❖ Pour the cold, aerated cream into the bowl of a spoon placed at the level of the top of the coffee. Continue until the cream is about 1cm deep.

❖ Remove the spoon and serve.

QUESTIONS

1. Name two places where beverages are served other than the restaurant.

2. What is meant by the 'responsible service of alcohol'?

3. If a customer appears to be getting drunk what should you do?

4. Why should abbreviations be avoided when taking the beverage order?

5. Why would guests normally prefer to order their wine after they have ordered their food?

6. When should the selected wine be opened?

7. How should the glassware be handled?

8. Why is a little wine served to the host before the other guests?

9. At what stage in the meal are liqueurs, brandies and ports offered?

10. List the equipment required for the service of tea.

11. What is a liqueur coffee?

Chapter 16

Espresso coffee

I'll have a decaf, skinny, weak latte …

On completion of this chapter you will have a basic understanding of:

❖ Espresso coffee styles

❖ How to organise and prepare a work area for making espresso coffee

❖ How to select, grind and extract the coffee

❖ The workings of an espresso machine

❖ The procedure for making an espresso coffee

❖ How to texture and pour the milk

❖ Presenting the coffee

❖ The importance of cleaning and maintaining an espresso coffee machine.

Australia now has a 'coffee culture', so that bars and restaurants, no less than coffee shops, are judged by the quality of their coffee. Waiters are often expected to act as **baristas**—that is, expert makers of espresso coffee.

Espresso coffee is made using a machine that forces water quickly through ground coffee to extract the drink we love. The espresso machine makes coffee with an intensely strong, concentrated flavour.

There are many different espresso machines available. The diagram on page 169 shows a typical one.

The Italian for espresso coffee is *caffè espresso*. The Italian word *espresso* has two related meanings: to be pressed out or extracted ('expressed' in English), and 'express' in the sense of fast or quick.

Espresso coffee styles

There are many different styles of espresso coffee. Most of them originated in Italy, but the Italian styles have been adapted and added to in other countries, including Australia. New styles become popular from time to time. The waiter must be able to provide advice on the styles of coffee available and serve them with the appropriate accompaniments. The principal styles of coffee popular in Australia are:

Espresso or short black

Long black

Cappuccino

❖ **Espresso**, or **short black**. An espresso, or short black, is a single coffee (30–35mL) served in a small cup (demitasse) or glass without milk. The coffee should be strong and hot, with a thick honey-coloured head, or **crema**. Many customers like their espresso sweet, so make sure that sugar is available.

❖ **Long black** (standard). The standard long black is milder than an espresso. It is made up of 30mL of espresso and 60mL, or two-thirds, hot water. It should have a thick head, or crema.

❖ **Strong long black**. A strong long black is made with 60mL of espresso (using a double filter holder) and 30mL of hot water. Like all black espresso coffees, it should have a thick head, or crema.

❖ **Cappuccino**. The cappuccino is made up of one-third (30mL) espresso, one-third hot

Caffè latte

milk and one-third foam. It should have a dense foam cap, which may be sprinkled with chocolate (optional). It is served in a cup, not a glass.

❖ **Mug cappuccino.** This is simply a double (60mL espresso) cappuccino served in a mug rather than a cup.

❖ **Caffè latte.** Caffè latte, usually served in a glass, is made up of one-third espresso, which is then filled with hot milk, with about 10mm of thick, creamy foam. (*Latte* is Italian for milk.) The milk should be approximately 70°C, not so hot that the glass cannot be held without a cloth.

Flat white

❖ **Flat white.** The flat white, served in a cup, is made up of one-third espresso and two-thirds hot milk. The only difference between it and the latte is that the flat white dispenses with the foam.

Macchiato

❖ **Macchiato.** The macchiato is a short (or sometimes long) black coffee 'stained' with a dash of milk or foam. Customers may prefer to add the dash of milk themselves, to taste. (*Macchiato* is Italian for spotted or stained.)

❖ **Ristretto.** The ristretto is an extra-strong, flavoursome coffee. It is made with the first 15–20mL of an espresso, extracted in 10–15 seconds, and served in a small cup or demitasse. (*Ristretto* is Italian for restricted or limited.)

Ristretto

❖ **'Weak' cappuccino, latte,** etc. Weak coffees may be made by using the ristretto technique (extracting the first 15–20mL of espresso only). To this, the ordinary amount of milk is added, diluting the coffee significantly compared with the standard 30mL of coffee versions.

Mocha

❖ **Mocha.** Mocha is made using 30mL espresso combined with one tablespoon of cocoa or chocolate syrup. It is then combined with hot milk and a fine layer of foam. It may be dusted with chocolate.

❖ **Iced coffee.** The iced coffee is 30–60mL of espresso added to chilled milk, with sugar syrup (optional), topped with aerated cream or ice cream and dusted with chocolate.

❖ **Vienna coffee.** A Vienna coffee is a long black espresso topped with whipped cream and dusted with chocolate. It may be served in a liqueur coffee glass.

Various other less popular styles of espresso coffee—affogato, corretto, doppio, caffè frappé, granita, etc.—are described in the glossary.

Organise and prepare work areas

Careful preparation (mise-en-place) is essential for the efficient service of good espresso coffee. This is the procedure:

❖ Turn on the machine. Most espresso machines take about 20 minutes to reach the operating pressure required to extract the coffee. A boiler pressure gauge on the machine will indicate when the correct steam pressure (1–1.3 bars) has been reached.

❖ Stack the saucers. Clean and unchipped saucers must be carefully stacked within easy reach of the operator. Don't stack them too high or they may topple over.

❖ Fill the bean hopper with coffee beans. The hopper of the grinder should be filled with fresh beans every day. Beans should not be left in the hopper overnight, so if the hopper has to be refilled you must be careful not to put more beans into it than will be used that day.

❖ Extra coffee beans must be available. For good coffee the beans must be freshly ground.

Espresso machine with cups and saucers stacked on top

Make sure that an adequate supply of fresh beans is always to hand. They come from the distributor in vacuum-sealed packs.

❖ Decaffeinated coffee should always be available. It may be in the form of beans or pre-ground.

❖ Different establishments make different use of cups and glasses for coffee. They may include thick glasses (used for lattes and long macchiatos), demitasse or espresso cups (used for short blacks or 'espressos'), double espresso cups (used for flat whites, long blacks and cappuccinos) and mugs (used for extra large coffees). Check every cup, mug and glass for chips, cracks, lipstick marks and coffee stains. Don't stack them too high as they can easily fall.

❖ Milk, foaming jug(s) and perhaps a thermometer are needed. Fresh milk should be kept in a refrigerator near the espresso machine. You must have full cream, light and soy milk available. You should also have a selection of different-sized stainless steel jugs for different types and amounts of milk. Less experienced baristas may use a thermometer to check the temperature of the milk.

❖ You will need plenty of clean cloths for wiping various things: the bench, the group handle, the steam wand (or 'arm'), etc. Keep separate cloths for different tasks.

❖ Some establishments serve a napkin with their coffee.

❖ Sugar may be served with the coffee or be available on the tables. Supplies must be checked. Sugar substitutes and diabetic sugar should also be available.

❖ Check that there is a full container of clean, dry teaspoons.

❖ Clear bench space. You will need a fairly large clear space next to the coffee machine. It must be wiped clean after making each coffee.

❖ Also needed is a container or drawer for used coffee. After the ground coffee has been used to make a cup of coffee, the remains must be knocked out of the group handle. This is done on the edge of a special container (sometimes a drawer or 'knock box'). The container must be emptied at the end of each shift and regularly washed out to avoid mould growth. Before a shift begins, check that the container for the used coffee is empty and clean.

Selection and care of coffee

Selecting a good blend of coffee, suitable for the espresso machine in use, and the proper care of the coffee once delivered, are important factors in whether or not the coffee you serve is good.

SELECTING THE COFFEE

The best coffee for espressos is usually a blend of beans from several parts of the world. The principal coffee companies source and roast their beans to create the consistent styles of coffee associated with their names. These styles vary considerably. An establishment's coffee supplier will give guidance on the styles of coffee to be chosen. This isn't just a matter of price or taste. Different blends suit different machines. Most coffee houses produce a premium blend for use with commercial espresso coffee machines.

STORAGE OF COFFEE

Coffee is a perishable commodity and exposure to moisture, light and air will cause it to deteriorate. Roasted coffee beans are delivered in vacuum-packed bags. They should be kept in a dark, cool place. Roasted beans should be used within five or six days after the bag has been opened.

Ground coffee very quickly loses its aroma and dries out. This is why the coffee beans are ground immediately before the coffee is made in the espresso machine, and why the grinder should be right next to the machine.

If there is ground coffee left in the grinder dispenser at the end of a shift, empty it into an airtight container and keep it for making iced or liqueur coffee, or use it to make the first few coffees at the beginning of a working day that 'season' the machine and are thrown away (see below). Don't keep coffee in a refrigerator as it will absorb other flavours and its taste will be tainted.

Grinding the coffee

The grinding process is crucial to good coffee. To produce high-quality espresso coffee you will need a commercial grinder with either conical or flat blades, which produces a fine grind.

THE COFFEE GRINDER

To make good coffee, coffee beans must be freshly ground not long before the coffee is made. There are different makes of grinders, which work in slightly different ways. The parts of a typical coffee grinder are shown in the diagram on page 167.

PREPARING THE GRINDER FOR USE

Before service begins, the grinder must be made ready:

❖ Check that the bean hopper and the dispensing chamber are clean.

❖ Half fill the hopper with fresh coffee beans.

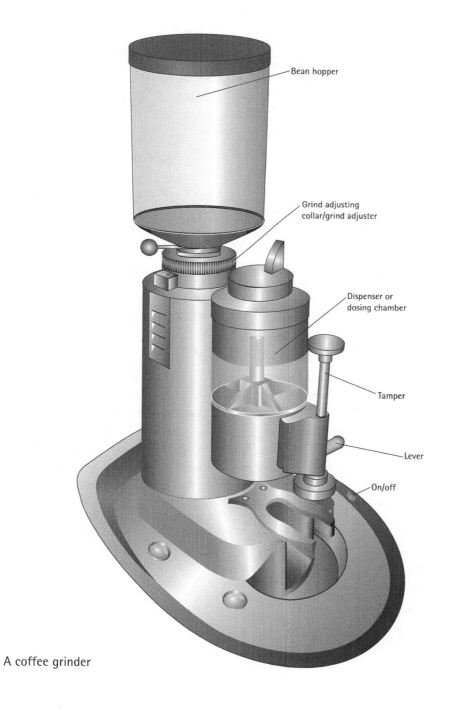

Bean hopper

Grind adjusting
collar/grind adjuster

Dispenser or
dosing chamber

Tamper

Lever

On/off

A coffee grinder

❖ Check that the machine is set for the correct grind size, and adjust the grind if necessary.

Conical grinding blades

ADJUSTING THE GRINDER

You must learn how to adjust the grinder to achieve the correct grind and dosage. This will require special instruction from the coffee supplier or your supervisor. The correct amount of coffee must be ground to the right degree of fineness and texture. While espresso coffee should be finely ground, it shouldn't be too powdery. It should have a slight grittiness. Only experienced staff should be responsible for adjusting the grind.

Flat grinding blades

The correct dosage for a single cup of coffee is 7–9g. The correct volume of espresso is 30–35mL. If the extraction of the coffee takes longer than 27–32 seconds, this indicates that the grind is too fine or the dose is too big; less than 23 seconds indicates that the grind is too coarse or the dose is too small.

The grind or dosage should be adjusted if it is taking more or less than 28–32 seconds to extract 30mL of coffee.

Besides the fact that different roasts require different timing and extraction, variations in grinding time may be caused by the environment, the condition of the blades or a badly maintained machine.

Adjustments to the grinder may have to be made during the course of a day. This is largely due to changes in moisture in the immediate environment. Coffee will absorb moisture or dry out, depending on the conditions around it. It will swell and grind more finely if there is moisture, and dry out and grind more coarsely if the atmosphere is dry. This can affect the rate of extraction (that is, the coffee will pour too quickly or too slowly) and the flavour will be affected. Small adjustments (finer or coarser) to the grind-adjusting collar should be made to re-establish the correct flow of coffee. In some work environments, the grinder may need regular adjustments.

The grind will also be affected if the grinder blades have become worn and need changing. You will know if the blades need to be replaced, as the coffee will feel warm, the motor may cut out, or it will become impossible to adjust the grind more finely without the blades touching. Grinders must be kept clean and serviced regularly.

The espresso machine

The principal parts of a typical espresso machine are shown in the following diagram.

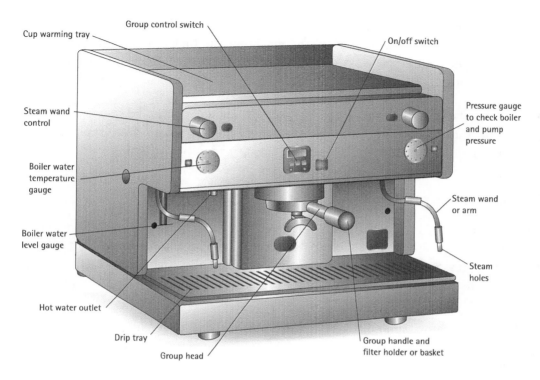

Cup warming tray

Group control switch

On/off switch

Steam wand control

Pressure gauge to check boiler and pump pressure

Boiler water temperature gauge

Steam wand or arm

Boiler water level gauge

Steam holes

Hot water outlet

Drip tray

Group head

Group handle and filter holder or basket

An espresso machine

PREPARING THE ESPRESSO MACHINE FOR USE

The espresso machine must be checked and prepared for use before coffee service begins, carefully following the manufacturer's instructions.

❖ Flush a little water through the group heads.

❖ Refresh the water left overnight in the boiler by draining off about three litres and replacing it with fresh water.

❖ Check the water level. The boiler should be about 65 per cent full.

❖ Check the pump pressure. In most machines, it should be 8–10 bars when extracting coffee.

❖ Check the boiler water temperature. In most machines, it should be about 100°C.

❖ Check the water temperature from the group head. It should be about 90°C.

❖ Release steam through the steam wand for about 30 seconds to blow out any vacuum that may have formed overnight.

❖ Before serving any customers, make two or three coffees through each group head and discard them. This will 'season' the machine. The first few coffees made will be unpleasantly bitter and sour-tasting. You may use any leftover ground coffee powder from the previous day for this purpose to avoid unnecessary waste.

Note that regularly during the working day you should check the water and pump pressure and the water temperature. Report any irregularities to your supervisor. No adjustments should be made by anyone who hasn't had special training.

Making an espresso

The espresso (short black) of 30–35mL of coffee is the basis of almost all the other coffee styles. Whether you are making a latte, flat white or cappuccino, your ability to extract the perfect espresso is crucial. It takes time and constant adjustment to achieve the right grind, the right temperature and the right steam pressure to make a first-class espresso. It also takes constant practice.

Different operators and different establishments will have different routines for making coffee, depending on the number of coffees to be made, the skill of the operator, the particular machines in use, etc. The procedure recommended below is not the only possible way of making good espresso.

PROCEDURE FOR MAKING AN ESPRESSO

❖ Select the appropriate cup or glass. These should be stacked on top of the espresso machine. This serves two purposes: they are easily accessible and they are kept warm. All espresso should be served in a warm cup or glass.

❖ Turn on the grinder and grind to fill the dosing chamber.

❖ Remove the group handle from the group head and knock out the used coffee. A maximum of two coffees can be made from each group head, so select the desired group handle (single or double measure). Knock out the spent coffee grounds and wipe out the filter basket with a clean dry cloth to remove any coffee particles. Don't clean the filter basket by flushing it with water.

❖ Dispense the ground coffee. While it is still grinding, place the group handle under the dispensing chamber. Flick the lever to release one dose of coffee (approx. 7g) or twice if making two espressos in the double group handle.

❖ Tamp the coffee. Before tamping, tap the side of the group handle with your hand or a manual tamper to disperse the coffee evenly. Using a manual tamper, or the tamper fixed to the grinder, pack (tamp) the coffee down gently. Keep the group handle as steady and upright as possible so that the coffee is packed down evenly. Now remove the group handle and wipe its rim with the palm of your hand to remove any loose coffee. Tamp the coffee again, this time applying more pressure. After tamping, the coffee should be evenly packed down with a flat surface.

❖ Clean the group head. Briefly run some hot water through the group head before you lock the group handle in place. This refreshes the group head and brings down the water temperature.

❖ Check pump pressure. The pump pressure gauge on most machines measures the pressure that is forced through the espresso, as well as the pressure of the water tank. The pump pressure should read 9–10 bars. This pressure is essential for creating the crema.

❖ Lock the group handle in place by placing it under the group head and turning it to the right.

❖ Turn on the machine, analyse pouring rate, start timing. After you have turned on the machine you have three seconds before coffee should start pouring into the cup. If the grind and dosage are correct it should take 25–30 seconds to extract a 30–35mL espresso.

❖ Check the quality of the coffee. The coffee should be deep brown in colour. The final product should have a honey-tinged layer of crema.

❖ Serve espresso quickly. Espresso should be served immediately after it has been made. After 90 seconds the crema will disappear. If you are making a number of coffees (lattes, flat whites) for one table, make the espresso last.

Texturing the milk

There are many ways to foam or texture milk, and the techniques vary from one establishment to another. The methods recommended here should ensure that your foam has the desired rich, thick, velvety texture.

THE MILK

Almost any milk—full-fat, low-fat or soy—can be textured to create a good foam, but most baristas prefer to work with full-fat milk because it is easier to achieve a perfect thick, velvety foam and it tastes better.

Non-fat milk 'fluffs' up easily, forming large bubbles if too much air is injected, and it separates easily, with dry foam on top and runny milk underneath, so it should be served immediately after steaming.

There can be problems with full-fat milk in Australia towards the end of summer and into autumn. The cows, which have calved in early spring, are no longer producing their best milk. The milk fat in their milk becomes liable to break down into fatty acids that inhibit the milk from frothing. Modified milks, specially designed for easy texturing, can be used to help overcome the problem.

The milk should be fresh and very cold. The colder the milk, the longer it takes to steam, and the longer it takes to steam the more smooth and velvety it will be when foamed.

THE JUG

Most operators prefer a stainless steel jug with straight sides, tapering in slightly towards the top, with a good pouring spout. Jugs are available in various shapes and sizes.

It is important to select a jug of the right size. Half-filled jugs give the best results. Some espresso machines don't produce enough steam pressure to foam or texture a full one-litre jug of milk.

Jugs should be stored in the refrigerator so that the cold milk poured into them is kept as cold as possible before it is textured.

STEAM PRESSURE

Steam creates a whirlpool effect. If the pressure is too much, the milk will be blown out of the jug; if it is too little, the milk won't spin properly. The steam pressure must therefore be correctly regulated.

PROCEDURE FOR TEXTURING THE MILK

❖ Select the correct size jug for the order (see above).

❖ Half fill the chilled jug with fresh cold milk.

❖ Turn on the steam wand and expel excess water and any milky residue and wipe.

❖ Place the tip of the steam wand in the milk.

❖ Turn on steam to full power.

❖ As soon as power is on, slowly raise the arm so that the tip is just below the surface of the milk and in the centre of the jug.

❖ As the milk expands (milk level rises), slowly move the jug down so that the steam wand remains just below the surface. This is known as 'stretching' the milk. Don't jiggle the jug up and down.

❖ The milk will have reached about 60°C, the point at which the jug becomes too hot to touch. The milk temperature will continue to rise to about 75°C, the correct temperature for serving. Place the palm of your free hand at the bottom of the jug to check the temperature of the milk. As soon as it becomes too hot to touch, turn the steam off before removing the steam wand from the milk.

❖ Remove the jug and wipe the steam arm with a clean damp cloth.

❖ Swirl the milk by rotating the jug for 15–20 seconds. If bubbles are visible, tap the jug on the counter several times.

❖ Pour the milk on to the espresso coffee, as appropriate for the type of coffee requested.

❖ Don't reheat the milk; use only enough milk for the job at hand.

POURING THE MILK

Steaming, or texturing, the milk is done in the same way for lattes, cappuccinos and flat whites, but the milk for each of these types of coffee is poured in a different way.

Pouring a cappuccino

After steaming the milk, keep it rolling in the jug, as recommended above. This will help produce the dense, meringue-like foam required. Rest the jug on the rim of the cup and pour steadily and quite fast into the centre of the cup. The milk will separate once in the cup, producing the one-third coffee, one-third running milk and one-third foam proportions required for a cappuccino.

❖ Alternatively, you can let the jug stand for 20 seconds, allowing the foam to settle on top. Using a spoon, hold back the foam and pour in the milk up to two-thirds full and then spoon the foam on top.

❖ The first of these methods is generally preferred. It is quicker, looks and tastes better, and the milk can be used for all three milk coffee types.

❖ Dust with chocolate (optional) and serve.

Pouring a flat white or latte

Flat whites and lattes should have less foam than a cappuccino. To achieve this, pour the milk from about 6cm above the cup, and pour more slowly than for a cappuccino. This will cause the foam to break down more (into runny milk) with the result that there will be less foam on top.

A latte may have a little more foam than a flat white. You can achieve this by holding the jug a little closer to the cup and pouring a little quicker than for a flat white.

Guest preferences and orders

Tastes in coffee change over time and from one guest to another. It is the waiter's job to serve coffee in the way guests prefer it. Some factors to be considered are:

❖ *Coffee strength.* Guests may ask for a strong or a weak coffee. A coffee correctly extracted (30–35mL in 27–32 seconds) will be a strong coffee. If the extraction takes longer than 30 seconds, or you end up with more than 35mL of coffee, the coffee becomes over-extracted and bitter because it is basically water running through at this stage. To make a weak coffee correctly, you should make a ristretto (15–20mL extracted in 10–15 seconds) and add milk or hot water to taste.

❖ *Cups, mugs and glasses.* Guests may choose to drink a latte in a cup, a cappuccino in a mug or a short black in a glass. Whatever the guest wants is 'right'.

❖ *Sugar.* Sugar comes in various forms and packages—white, brown, crystal, tablet, even liquid. No establishment will be able to provide a complete range, but sugar substitutes, such as saccharine, should be available—for example, for guests who suffer from diabetes.

❖ *Milk.* Most establishments will stock low-fat and soy milk in addition to regular milk. A guest may ask for a jug of hot or cold milk on the side.

❖ *Decaffeinated coffee* is increasingly popular. Decaffeinated coffee beans and decaf ground coffee are available. Decaffeinated coffee is made in the same way as any other espresso coffee.

With so many variations possible, you must be able to understand and note exactly what customers have asked for: 'a cappuccino with no chocolate on top', 'a decaf skinny latte', 'a short macchiato with cold milk on the side', 'a weak soy milk flat white', and so on.

Presenting the coffee

Having prepared a great coffee, it is essential that the presentation add to the guest's pleasure and appreciation of the product, as with all other food and beverages. The following are the key considerations for an effective and appropriate presentation:

❖ the use of clean (stain-free) white ceramic cups or glasses

❖ checking for drips and spills prior to serving

❖ offering a folded napkin alongside the coffee (as per house standard)

❖ ensuring that a clean, dry teaspoon always accompanies the coffee (the teaspoon is placed on the right-hand side of the saucer, either under the handle or across the top)

❖ placing the coffee with the handle facing to the guest's right

❖ serving a glass of cold water with the coffee, offering to bring water, or showing the guests where they may help themselves to water, according to the establishment's practice.

Cleaning the espresso machine and grinder

Frequent and thorough cleaning and proper maintenance of the espresso machine and the grinder are absolutely crucial to the making of good coffee. If the machines aren't clean, the coffee will lose its flavour and begin to taste rancid. Coffee beans contain oil, which will quickly go rancid and ruin the coffee if not cleaned in time.

Espresso machines must be cleaned daily to avoid the build-up of bean oil and coffee grounds. You must also clean the steam wand with a clean sponge or cloth after each use and flush the steam holes, clearing them of milk residue. The machine must be wiped frequently to remove milk and coffee splashes.

Most parts of the machine need a good clean, scrub or wipe-over, using both wet and dry cleaning methods. It is best to use a clean sponge and tea towel. Scourers shouldn't be used.

Every espresso machine comes from the manufacturer with an operating manual that includes instructions on how it should be cleaned and maintained. It is very important to follow these instructions. If you are in doubt, ask your supervisor or coffee distributor for advice. Cleaning procedures will be required during the working day and at the end of the day's work. The espresso machine and the coffee grinder should be checked for wear and tear at least weekly.

CLEANING THE MACHINE DURING THE WORKING DAY

Frequently, during the day:

❖ check that the water coming from the group head is clean and clear

❖ flush out any water that has built up in the group heads

❖ expel excess water from the steam wands

❖ wipe the steam arm after each use with a clean damp cloth.

CLEANING AT THE END OF EACH DAY

At the end of each working day:

❖ wipe the cup tray and the panels of the machine with a clean, soft cloth

❖ clean the steam wands by wiping them down with a clean, soft cloth; make sure that the steam holes are clear; remove any build-up of milk from them, using a paperclip if necessary

❖ remove the drip tray, wash and dry it, and replace it

❖ remove the group handle baskets from the group handles and clean them

❖ soak the group handles in hot water to remove oily deposits

❖ back-flush the machine using a 'blind filter'—that is, one with no holes in it—to build up the pressure before flushing, carefully following the manufacturer's instructions.

If the machine is very busy, these procedures, including back-flushing, may have to be repeated more than once during the working day.

CLEANING THE GRINDER

At the end of the working day:

❖ remove any coffee beans remaining in the bean hopper and store them correctly

❖ wipe the bean hopper with a clean dry cloth

❖ brush the top of the grinder blades to remove coffee grains resting on top of them

❖ remove any leftover coffee from the dispenser and brush it clean.

QUESTIONS

1. How much coffee is there in an espresso?

2. What is the difference between a latte and a flat white?

3. What is a macchiato?

4. Where should clean coffee cups be stacked and why there?

5. Why should the coffee beans be ground just before the coffee is made?

6. What is the crema on an espresso?

7. Why should the milk and the jug it is foamed in be kept very cold?

8. How do you achieve more or less foam when pouring the milk into a cappuccino or a latte?

9. How should a cappuccino be presented to a guest?

10. How frequently must the parts of an espresso machine be cleaned?

Chapter 17
End-of-service procedures

First-rate meal, well served, but if we could only find a waiter to give us our bill we would be able to leave. The show begins in ten minutes.

When you have read this chapter you will have a basic understanding of the correct ways to deal with:

❖ Preparing and presenting a bill

❖ Payment procedures and methods

❖ Tips (gratuities)

❖ Farewelling guests

❖ Tidying, cleaning and resetting after service.

We have already stressed the importance of greeting and receiving guests warmly and professionally and ensuring that effective and efficient meal service takes place. However, the total experience for the guest doesn't end there; an equal emphasis must be placed on the end-of-service procedures.

Preparing and presenting a bill

The methods by which bills are prepared and processed vary from one establishment to another. They range from handwritten dockets to high-tech computerised systems. The two purposes of a guest's bill are to inform the guest of the amount to be paid (giving details of what is charged for) and to act as a control system for the establishment. To ensure their control systems operate effectively, establishments must make certain that their new service staff understand how the billing system works before they begin to serve guests.

Guests' bills may be presented at the table, at the bar or at a cashier's desk. No matter where it is presented, the bill should be kept up to date at all times. Where possible, the bill should be kept ready for presentation as soon as the guest requires it. This may not always be possible, particularly when beverages are being served right up to the time of the guests' departure.

PRESENTING THE BILL

It is essential that you are alert to signs that guests may want their bill. Nothing is more irritating to guests than to be kept waiting while they try to attract the attention of a waiter to ask for their bill (or 'check', as the Americans call it). This is particularly so for busy businesspeople at lunchtime. Many a promising restaurant has failed because it earned a reputation for slow service, and the fatal slowness may well have been in the bringing of the bill rather than in the actual food service.

Generally speaking, bills should not be presented until they are asked for, but some establishments that specialise in quick service and a high turnover of guests place the bill on the table before the end of the meal.

When a bill is presented at the table it is placed in front of the host (probably the person who has asked for the bill) on a small plate from the right. Either the bill is folded so that the amount to be paid cannot be seen by the other guests, or it is placed in a **billfold** that serves the same purpose. If there is no obvious host, you may place the bill in the centre of the table.

Bills presented at bars should be presented on a plate, folded or in a billfold.

If the establishment requires guests to pay at a cashier's desk as they are leaving, make this clear to the guests to avoid confusion and delay.

Don't hover around waiting for your guests to pay; leave them alone to pay in their own time. Remain alert, though, so that when they have paid (or signed) for their meal, there is no unnecessary delay while they are kept waiting for you to collect the payment.

Methods and procedures for payment

Common payment methods include cash, credit cards, the electronic funds transfer at point of sale (EFTPOS) system, vouchers and charge accounts. You must be familiar with the procedures for these various methods of payment and know which methods of payment are acceptable to the establishment.

Cash payments are very simple, requiring only the settling of the bill and the tendering of the guest's change.

Credit cards. The precise procedures for use with credit cards will vary from establishment to establishment. When the card is placed on the bill you should collect it and, before processing it, check:

❖ that the establishment accepts the kind of card presented

❖ if the charge is to be billed to a credit or savings account

❖ its expiry date

❖ that it has been signed.

Your employer will advise you of the limits that apply to credit card sales and if a processing fee is applicable to the card. It is also wise to check the number against the current warning bulletin.

Processing credit cards involves either the manual completion of a slip for signature by the guest, or for the card to be run through an electronic swipe that will produce a transaction slip for the guest to sign. When slips are prepared, it is usual not to complete the 'Total' line so that the guests can confirm the amount signed for and add a tip if they so wish (see below).

After you have checked that the guest has clearly completed the 'Total' line on the slip with the sum due (or more), and you have checked that the signature on the slip is the same as that on the card, the card should be returned to the guest along with the guest's copy of the slip. The guest's slip should be returned either upside down or (better) in the billfold that was used to present the bill.

EFTPOS. The electronic funds transfer point of sale system is commonly available throughout the industry. Customers can now pay through EFTPOS using a credit or debit card. To commence an EFTPOS transaction, the customer's card is swiped through a terminal. Once the card is swiped, the display on the terminal will guide you through the process. The hardware used to process a transaction may vary from one establishment to another and, with this in mind, it is a normal process that employees undertake in-house training to familiarise them with the specific equipment and procedures used at that establishment.

Vouchers may be used for pre-paid transactions. They may be gift vouchers, for example, or part of a package, or through a complimentary offer. The voucher takes the place of money payment, but it must be accounted for like any other form of payment.

Charge accounts. Before a guest can be allowed to charge the cost of a meal (or drink) to a charge account, the transaction must have been authorised by management. It may be necessary to check the guest's signature against a charge record or, if it is in a hotel, the guest's name against a room number. Procedures for recording information on charge account transactions vary from establishment to establishment. You must, of course, strictly follow house procedures.

Tips (gratuities)

Tips are a bonus for especially satisfying service and shouldn't be thought of as a right. There are no hard-and-fast rules on the amount of the tip or, indeed, on whether one is left or not. Never expect or anticipate a tip. If the guest has paid in cash, always place the full change on the table unless the guest has very clearly indicated that the excess payment is intended as a gratuity.

Never allow guests to leave with the humiliating impression that you think they should have tipped you, or that their tip could have been more generous. If you do, they will not come back. Remember, the waiter has absolutely no right to a tip.

In some establishments it is house policy that all tips should be pooled and distributed among all the staff. This is called the 'trunk system'. Where this system is in place, it must be very strictly adhered to or there will be bitter disputes. In other establishments, individual waiters keep the tips they have been given.

Farewelling the guests

The last impression guests are given as they leave after a meal is as important as their first impression on arrival. The farewell should be warm and friendly, and as personal as possible. If you are not too busy serving other guests, assist those departing by moving their chairs for them, collecting their personal belongings (not forgetting coats, hats and BYO bags!) and offering to call for a taxi.

If it is not physically possible to assist your guests to leave because you are busy serving others, at least acknowledge their departure with a nod and a smile. If you can, wish them 'Good evening' and thank them for coming. You should have taken the trouble to note their names (the credit card is an invaluable reminder for names); if you know the name, use it: 'Goodnight, Mrs Hill. We look forward to seeing you again soon.'

Tidying, clearing and resetting

The waiter's responsibilities don't end with the departure of the guests. When the guests have left, the tables and service areas must be cleared of used and soiled items and the tables prepared for use again.

The procedure for resetting the tables and work areas will vary from one establishment to another. In some establishments, each table is reset for the same meal service as soon as the guests sitting at that table have left. This allows the table to be used again, increasing the number of covers served in that meal service. In other less hurried establishments, after the guests from one meal service have left, the tables are reset for the next service—for example, when one set of guests has finished their lunch the table may be reset for dinner.

PROCEDURE FOR CLEARING AND RESETTING

❖ Remove coffee cups and centre items, and glassware. (If you have kept the table tidy throughout the service of the meal, these should be the only items left on the table when the guests depart.) The cups and saucers should be carried using either the two- or the three-plate carrying technique (see Chapter 7). Don't stack the cups. Glassware should be removed on a drinks tray. The remaining centre items are removed by hand.

❖ If tablecloths are used in the establishment, the table must be reclothed. If the table is reclothed after service (when the restaurant is empty), use the clothing procedure described in 'Clothing procedure' in Chapter 4. If the table is reclothed during service, follow the procedure described in 'Changing a cloth during service' in that chapter. If tablecloths are not used, all tables must be carefully wiped down.

❖ Whether the table is reset or not, ensure that all the chairs are returned to their original positions around the table. Don't forget to check the chairs for crumbs.

❖ The procedures for setting or resetting covers are described in Chapter 4.

❖ In most establishments, workstations are restocked with cleaned, polished equipment immediately after the completion of service, in preparation for the next service (see 'Station mise-en-place', in Chapter 4).

QUESTIONS

1. What are the two purposes of a bill?

2. Why is it important to present the bill promptly?

3. What should you do with the bill if it is not clear who is the host?

4. What is the procedure for accepting payment by credit card?

5. Why should you never stand and wait for a tip?

6. How can you make a good impression when farewelling guests?

7. You should make sure that guests take their personal belongings with them when they leave. What items are they particularly likely to forget?

8. How should you carry cups and saucers when clearing a table?

9. How should workstations be left?

Chapter 18

Room service

Procedures are put in place to ensure we respect the guest's privacy and avoid embarrassing situations like this. Always remember: knock three times and announce yourself.

On completion of this chapter you will have a basic understanding of:

❖ The preparation of equipment and food and beverage items for room service

❖ The correct way to take and process room service orders

❖ The setting up of trays and trolleys for all types of service

❖ How to enter a guest's room

❖ The presentation of room service meals and beverages

❖ The presentation of room service accounts

❖ Room and floor clearing procedures.

Room service is the service of food and beverages in guests' rooms in hotels or other accommodation establishments, such as motels or serviced apartments. In all-suite hotels it is often referred to as 'in-suite service'.

In establishments of any size, there is usually a specialist Room Service department responsible to the Food and Beverage Manager. The Room Service department must work closely with the Kitchen, Front Office and Housekeeping departments to make sure that the standard of service satisfies, or more than satisfies, guests' expectations. Hotels are often judged, as much as anything else, by the standard of the room service they provide. A five-star property will be expected to provide room service for at least 18 hours of the day, if not all hours of the day and night, and that service must at all times be friendly, quick and efficient.

Preparing room service items for service periods

Most modern hotels have a single central pantry for the Room Service department located conveniently near the kitchen and the service lift. This pantry should be fully equipped for quick and efficient service to the rooms.

Room service catering can involve the delivery of everything from complimentary items and items for which no charge is made (such as ice buckets and glasses) through drinks or light snacks to full à la carte meals with wine.

The items available for service in rooms will normally be listed on a special room service menu, but in a superior hotel guests will expect any reasonable request to be met.

There may be different sections within the room service menu listing the items available at different times of the day—for example:

❖ Breakfast: 6 am to 11 am

❖ All-day dining: 11 am to 11 pm

❖ A la carte: 7 pm to 10.30 pm

❖ Night owl menu: 11 pm to 6 am

The pantry must be stocked with sufficient equipment to ensure that all orders can be met promptly even at the busiest times. A typical hotel might pride itself on meeting all room service orders in less than 30 minutes of the order being taken.

To perform efficiently, even at the busiest times, a Room Service department must analyse demand to allow for occupancy levels and special circumstances (for example, the delegates at a major convention all requiring an early breakfast before a day's outing). Forward planning should provide the necessary staffing levels and ensure that the right number of trays and trolleys are prepared for service.

Storage of equipment and products

Room service equipment includes such items as:

❖ trays and trolleys

❖ cutlery, crockery, linen and glassware

❖ selected food and beverage items

❖ printed materials.

All these items must be stored in a safe, hygienic, orderly and accessible manner.

Every item should have a 'par stock' level. If items are kept in the pantry at their par stock level, there should be sufficient to meet demand but not too many or too much. The storage places for each item should be clearly labelled and its par stock noted so that any deficiencies are immediately obvious.

Items must also be stored safely to reduce the risk of accidents and breakages. Accidents and breakages are not only dangerous and expensive, but may impact badly on guest service.

Preparing room service equipment

All well-managed establishments will have their own standard policies and procedures to ensure a consistent standard of room service. Establishment procedures must, of course, be followed by room service staff.

Different tray and trolley set-ups are dictated by the menu items to be served and will also include provision for common requests for items not included on the room service menu. The details of advance tray or trolley set-up will vary from establishment to establishment, but in most instances there will be standard set-ups for:

❖ tea and coffee trays

❖ ice buckets

❖ breakfast trays and trolleys

❖ snack trays

❖ dinner trays or trolleys

❖ champagne or wine trays

❖ fruit basket trays

❖ butters

❖ condiments

❖ bread baskets

❖ hot boxes.

Taking room service orders

Most room service orders are given by telephone. The telephone is therefore the first point of contact with room service staff and so good telephone technique is vital in creating that all-important favourable first impression.

The person answering the telephone must have a good knowledge of the menu and a professional telephone manner (see Chapter 4).

The telephone must be answered quickly. The benchmark for a five-star hotel is no more than three rings before it is answered. Then pay special attention to:

❖   the greeting

❖   introduction of department and self

❖   use of the guest's name.

This can be achieved by an answer along these lines:

'Good morning, Mr Stephens. This is Room Service, Mark speaking. May I help you?'

The use of the guest's name has the advantage of ensuring that the items requested are delivered to the right guest—and charged to the right account—as well as making the guest feel known and valued. Most properties now have advanced telephone systems with a room number and a name display, but if these are not available you should have a current rooming list handy. Continue to use the guest's surname at all stages of room service.

After greeting the guest and confirming his or her name, continue as follows:

❖   Write the order down carefully on an order docket as you speak to the guest. Don't forget to record the room number—this is often forgotten!

❖   Always seek opportunities to 'up-sell'—that is, to use suggestive selling techniques to increase the value of the order (see Chapter 6).

❖   Be flexible and helpful if the guest requests items not on the menu.

❖   Repeat the order to the guest, clarifying any doubtful details.

❖   Tell the guest approximately how long it will take for the order to be delivered. It shouldn't be more than 30 minutes.

❖   Check that all the details (including the room number) are correctly recorded on the docket. Include the time that the order was taken.

❖ Enter the order in POS (see Chapter 6).

❖ Promptly distribute the order to the appropriate personnel, both in the Room Service department and in other departments if necessary, most obviously the kitchen.

Not all room service orders are taken over the phone. Breakfast orders, for example, are frequently in the form of doorknob dockets completed by guests before they go to bed and left on their doors for collection. These orders must be checked, clarified if necessary and correctly distributed.

Setting up trays and trolleys

Once the order has been distributed, a suitable preset tray or trolley should be selected and set up appropriately depending on the number of covers, the food and beverage items ordered, and the meal or snack requested. There should be enterprise standards for tray or trolley preparation for the different meals (breakfast, lunch, dinner) and for complimentary items and special requests.

All service equipment should be checked to make sure that it is clean and undamaged. Trolleys should be checked to ensure that they move properly.

Set-ups should be checked to make sure that the presentation is attractive, and that they are well balanced and safe. Ensure that the necessary condiments are included in the set-up.

Collection of the orders

When the trays or trolleys have been checked to make sure that they are correctly set up, food and beverage items (when ready) should be collected promptly and in the right order, with the appropriate accompaniments.

Food and beverage items should be checked, with attention to such details as:

❖ food and beverage temperatures

❖ portion sizes

❖ visual presentation as per recipe standards

❖ wine details, including vintages.

Food temperatures must be maintained from the time the food is collected to the time it is delivered to the guest. Plate covers, food warmers and/or hot boxes should be used to keep food at the right temperature.

Collect the guest's account and confirm that it matches the order. The account must be taken to the guest's room along with the items ordered.

When all items have been checked—service equipment, food and beverages as ordered, and the account—they should be taken to the guest's room without delay.

Entering a guest room

Respect for a guest's privacy is the primary consideration when entering a room. There will probably be particular establishment procedures for this, as for all other aspects of room service, but the following will usually apply:

❖ Approach the room quietly.

❖ Knock firmly and say 'Room Service' clearly and confidently, remembering that your voice must carry through a closed door.

❖ Listen for the guest's response and react accordingly, waiting outside or entering the room. If there is no response, knock and announce 'Room Service' again. Don't go in until the guest opens the door or you have been asked to enter.

❖ When you have entered, address the guest by name—'Good morning, Ms Tan. Here is your breakfast'—or whatever is appropriate.

Continue to use the guest's surname while making polite conversation throughout the room service procedure.

Presentation of room service food and beverages

Exactly where trays are placed and trolleys set up will vary according to circumstances, depending on the equipment being used, the design of the room, the position of the furniture and the guest's particular wishes. There will be appropriate enterprise procedures that should be followed, subject to the wishes of the guest.

❖ Confirm that the tray or trolley is being placed where the guest wants it.

❖ Set them up where directed, bearing safety in mind. Advise the guest of any potential hazard—for example, the hot box or the coffee pot may be too hot to touch.

❖ Position the furniture appropriately.

❖ Light a candle, if applicable.

❖ Explain the contents of the tray or trolley.

❖ Serve the food and beverages, following company procedures.

❖ Ask the guest whether anything else is required.

❖ Present the account for signature.

❖ Explain the clearing procedures. Typically, guests may be requested to put trays or trolleys outside their room when they have finished.

❖ Farewell the guest in a friendly but courteous manner and leave the room quietly.

❖ Check the floor or passage outside the room and remove any used trays or trolleys that may have been put there.

❖ Present the signed charge account to the cashier or other correct department, following establishment procedures.

Clearing room service areas

As noted, trays and trolleys will usually be placed outside the rooms by guests. It is important that they are cleared promptly, as messy and untidy floors or passages reflect very badly on the establishment. Good working relations and good communication channels between the Housekeeping and Room Service departments must be maintained to ensure that floors are cleared quickly. This requires a combination of good systems and procedures with the common sense and initiative of individual staff members.

Different establishments will have slightly different procedures; however, typically, the procedure will be for the Room Service order-taker to record the following on a despatch sheet or checklist:

❖ date

❖ room number

❖ whether tray or trolley taken (an agreed code may be used, such as T for trolley and O for tray)

❖ time the order was taken

❖ time the order was delivered

❖ person delivering the order

❖ time the order was cleared.

The Room Service order-taker will usually be the person responsible for controlling floor service procedures and for directing staff to clear rooms and floors. In addition, there should be regular floor checks, at least hourly, to ensure that used trays and trolleys and miscellaneous items are quickly removed. Housekeeping staff must notify Room Service if and when used items need to be cleared from rooms or floors, so close coordination between Room Service and Housekeeping is essential.

When a room or floor has been cleared after room service, the Room Service order-taker must be informed.

Floors must be cleared quickly and quietly, but while doing so staff must take care that equipment is securely placed so that it can be moved safely.

Once cleared from the floors, unconsumed food and beverages, food service equipment, trays and trolleys must be returned to Room Service via the service lift.

Cleaning and storage

Trays and trolleys should be taken to the wash-up area. Waste items will be removed and food service equipment will be washed.

Trays and trolleys must be cleaned and dismantled safely and hygienically following enterprise procedures. These will include procedures for:

❖ stacking equipment

- ❖ dirty linen

- ❖ storing usable items.

Clean trays and properly dismantled trolleys must be correctly replaced with other usable items in the Room Service pantry.

Stock of all items should be checked against par stock and requisitioned for restocking as required to meet the establishment's standards.

QUESTIONS

1. What is room service?

2. When will room service be available in a superior hotel?

3. What is meant by the 'par stock' for room service equipment?

4. Why should room-service staff use guests' names when speaking to them?

5. What is the procedure for taking a room service order?

6. What are the 'necessary condiments' for a breakfast of corn flakes, eggs and bacon, toast and marmalade, and coffee?

7. How should room service staff enter a guest's room?

8. Suggest some hazards about which guests may need to be warned when a room service meal is presented.

9. With what other departments in a hotel will the Room Service department most frequently need to communicate?

10. When used room service equipment has been cleaned and put away it should be 'checked against par stock and requisitioned for restocking as required'. What does this mean? What do you think the procedure is likely to be?

Chapter 19

Function operations

How often do we attend a function where the standard of service is not only unsatisfactory for the guests but an embarrassment to the staff employed to provide it?

Far too often function staff are left to survive as best they can when prior knowledge of the simplest of skills would have given them a sense of pride, and the service expectations of the guests would have been fulfilled.

When you have completed this chapter you will have a basic understanding of:

❖ Styles of function

❖ The variation of function covers

❖ Food and beverage service in function operations

❖ Function staff organisation.

Functions offer people who lack formal training their greatest opportunity to gain part-time employment and establish themselves in the hospitality industry. The relatively simple skills required of a waiter for a specific function can easily be demonstrated by the employer, and working at the function will ensure that those skills are practised repeatedly over a short period.

Styles of function

Function catering can involve anything from the simple service of sandwiches and coffee or tea to gala banquets, and the functions can take place anywhere, indoors or out-of-doors, from the garden of a private home to a grand ballroom.

The precise tasks required of service staff at functions requiring food and beverage service are defined by:

❖   the client's needs

❖   the occasion

❖   the types of food and beverages to be served

❖   the amount the client is prepared to spend.

Within these constraints the only limits to the diversity of function operations and the services provided are the physical facilities available and the imagination of the caterer.

Function preparation

When preparations are being made for a function, it is important that every member of staff should be fully informed about the requirements of the client—the host or organiser of the event. This information is recorded on a document, often called a *function sheet*.

The responsible person in the establishment where the function is to be held (in a large hotel, usually a banquet sales representative) communicates directly with the client and establishes what is wanted. Once agreed, the details of the function are carefully recorded and confirmed by both parties. The function sheet (which may be known by some other name) can then be prepared.

The information normally detailed on the function sheet includes:

❖   the name of the client

❖   client's contact person

❖   type of function

❖   date and time of function

❖   guaranteed number of guests

❖   room allocation

❖   floor plan

❖   menu and beverage list

❖   timing schedule for the service of food and beverages, and guest activities (speeches, presentations, etc.)

❖ specially requested items or services (car parking, décor, sound systems, music, entertainment, etc.)

❖ price and billing procedures

❖ establishment's contact person.

In sufficient time before the date of the function, the function sheet is sent to the departments that will contribute to the set-up and operation of the function. The various departments must work together closely to ensure that all the details of the function are fulfilled as requested.

Someone on the establishment's staff will have special responsibility for supervising the staff involved in the operation of a function. This person may be called the *function supervisor*. The function supervisor will allocate duties to members of staff, making sure that all details of the function, as listed on the function sheet, have been catered for, and that the staff involved know exactly what is expected of them.

It is essential that every member of the function staff (whether permanent or casual) is aware of, and can apply, the establishment's standard operating procedures. These will ensure consistency, hygiene, safety and efficiency in the setting up for, and service at, the function.

Setting up for a function includes the following tasks:

❖ the set-up of tables, chairs and other large equipment as required

❖ the mise-en-place of the small equipment (linen, crockery, cutlery, glassware, condiments, etc.)

❖ the setting of covers

❖ the setting of service areas

❖ preparations for the special requirements of the client as indicated on the function sheet.

Function covers

The cover for a function is dictated by the menu items to be served. Functions usually have set menus and the cutlery items are set in the order in which they will be used (see Chapter 4).

DESSERT COVERS

Dessert cutlery presents special problems for which there is more than one acceptable solution. At functions it is very likely that there will be pressure on space on the table, which adds to the advantages of bringing the dessert cutlery to the table only when it is needed for the sweet course (see Chapter 4). If the dessert cutlery has been laid with the original cover, it will have to be corrected. For the procedure for correcting dessert covers, see Chapter 6.

A possible function cover

Set menu cover with dessert gear

GLASSWARE

At functions there may not be room to place all the glasses in a single line in order of their probable use. If there are more than two glasses, they may have to be arranged in a triangle (see Chapter 4).

CUPS AND SAUCERS

If cups and saucers are required when the cover is laid, they are placed above and slightly to the right of the main knife, with the handle of the cup turned to the right so that it may be conveniently grasped by the guest without any need to turn it. Alternatively (and more usually for dinner functions), cups and saucers may be placed immediately before they are required for use (see Chapter 15).

Function service

The basic styles of service offered at functions are no different from those in restaurants—plate service, silver service, etc. The skills required for each style of service are described in previous chapters (see Chapter 7 for plate service and Chapter 8 for silver service).

FUNCTION SERVICE SKILLS

The basic skills required for function service are:

❖ providing hospitality (see Chapter 1)

❖   setting tables (Chapter 4)

❖   carrying a tray or platter (Chapters 5, 13, 15)

❖   use of a service cloth (Chapter 7)

❖   carrying or clearing plates (Chapters 7, 9)

❖   pouring wine (Chapter 15).

FUNCTION STAFF ORGANISATION

The number and organisation of the staff at a function will depend on the particular requirements of different functions. Different styles of

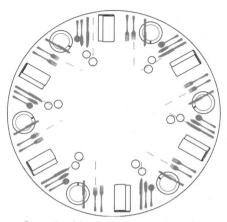

Round table setting for a function

functions demand different staff arrangements and procedures. In some circumstances, the waiting staff work as a team serving all the guests at a number of tables; in others, they have particular stations allocated to them and they are responsible for serving a set number of guests.

In most circumstances, the team organisation best satisfies the needs and expectations of the guests and lightens the load on individual serving staff.

TEAM OPERATIONS

At large functions, to preserve a smooth and quick level of service, the service staff is often divided into teams.

Each team is made up of the number of staff needed to handle the service of a complete table. For example, tables seating ten guests may require a team of five staff. The team of five is itself divided into two groups with different functions—a serving group of three and a 'running' group of two.

The runners are responsible for picking up plated items from the kitchen and transferring them to a service area within the dining-room. The serving group collects the plates (or other items) from the service area and serves them directly to the guests.

When one table has been served, the whole team—servers and runners—moves on to another table.

Clearing is done in the same way. The serving group clears the table and carries the used items to the service area, and the runners remove them from the service area to the washing-up area.

The number of tables a service team can handle depends on the complexity of the menu, the style of service and the facilities at the venue. A food service team of five, plate-serving an 'average' three-course dinner, might be expected to serve up to 20 tables of ten guests—that is, up to 200 guests.

BEVERAGE SERVICE AT FUNCTIONS

Responsibility for beverage service is normally prearranged by a station drinks waiter. The precise duties involved will depend on the selection of beverages and the way they are to be served. Duties may range from serving a variety of beverages from a tray to table service of apéritifs, sparkling wines, table wines, after-dinner drinks and coffee (see Chapter 15).

If food and beverage service at functions is to be coordinated, the beverages must be served alternately with the food. The beverages for each course would usually be served before that course.

At functions there will usually be separate teams responsible for serving the beverages, alternating at the tables with the food service teams. Even if there are not, and the same waiters are responsible for serving both food and beverages, the beverages and the food are not served simultaneously but separately, to avoid confusion.

QUESTIONS

1. What does the hospitality industry mean by 'a function'?

2. Write down six different styles of function that might need the services of waiters.

3. What is the purpose of the 'function sheet'?

4. What should a function cover consist of?

5. In what circumstances do function staff work in teams?

6. What is the job of the runners in a function team?

7. How many five-person teams do you think would be necessary to serve a standard three-course dinner for a function with 400 guests?

8. In what ways might correct drinks service be different at a large function compared with a dinner for four in a restaurant?

Glossary

This glossary is not simply an explanation of the technical words used in the text of *The Waiter's Handbook*. It also covers a large number of the terms used in food and beverage service that are frequently found in menus in the *descriptions* of the dishes on offer. It therefore covers much of the food product knowledge with which every waiter should be equipped.

The headwords are in **bold** type. After the headword, if it comes from a language other than English, its language of origin is given in brackets in abbreviated form in *italics*—for example, (*Fr.*) for French. Most of these are obvious enough, but some are a little rare—for example, (*Heb.*) for Hebrew or (*Yid.*) for Yiddish. Occasionally, the reference is to a culture rather than to a particular language—for example, (*Ind.*) for Indian (which is not the name of a language).

For words of non-English origin, especially the many French words, there follows in square brackets a simple pronunciation guide. This isn't as simple as it may seem because some languages make use of sounds for which there is no equivalent in English: for example, the nasal in in the commonly used French word *vin* (wine), which we have tried to indicate by using brackets: [vi(n)]. There is also the problem of the soft j sound in French, which we have indicated by –zj, as in *jus* [zjoo] (meat juice) or *aubergine* [oh-bear-zjeen] (eggplant).

All French nouns have a gender—masculine or feminine—as well as a number—singular or plural. Adjectives used to describe nouns have to agree with them in both gender and number, the adjective following the noun, as in *marrons glacés* (candied chestnuts). Most plurals are easy, just add s, but some are less obvious, as in *gâteaux* (cakes). When the plural isn't obvious, we have included it after the headword, therefore **gâteau(x)**.

Because French words are so common in food service, and because French adjectives have to agree with their nouns, we include the gender when a French noun is explained—for example, **crème** (*Fr.*) (*fem.*). Because *crème* is feminine, any qualifying adjective must also be feminine, as in **crème brûlée**.

As French adjectives have both masculine and feminine forms, we have included both when you are likely to encounter them. In most cases, the feminine form is made by adding **e** to the masculine form. We have indicated this by putting the additional **e** for the feminine form in brackets—for example, **gratiné(e)**.

Most words used in definitions that are themselves defined in this glossary are printed in **bold** type.

abalone [aba-lony] Large **mollusc** (genus *Haliotis*) in rough ear-shaped shell.

abalone mushroom Oyster mushroom; large white mushroom with a slight taste of **seafood**.

abats (*Fr.*) (masc. pl.) [a-ba] **Offal** (always referred to in the plural—les *abats*). (*See also* **sweetbreads**.)

aboyeur/se (*Fr.*) [a-boy-ur/urze] The person in a traditional kitchen **brigade** who controls the hotplate and is responsible for communication between kitchen and waiting staff, and who 'calls up' the orders. (Literally the 'barker', from *aboyer*, to bark.)

accompaniment Condiment offered by the waiter to add relish to the dishes served. Many condiments are traditional, e.g. mint jelly with lamb or **parmesan** cheese with **pasta**.

achar (*Ind.*) [a-chahr] **Pickle** of fruit or vegetables **marinated** in oil. Eggs, meat and fish can also be preserved in this way.

acidity Wine-tasting term indicating tartness.

advocaat Liqueur of egg yolks, sugar and brandy; egg nog. (In Dutch, 'advocate' or 'lawyer'.)

affogato (*Ital.*) [affoh-gahto] Drink or dessert of espresso coffee blended with ice cream, or served separately for the guest to mix. A shot of liqueur can be added if required. (Literally 'drowned'.) (*Compare* **tartuffo**.)

aflatoxin Food poison (toxin) derived from the yellow mould *Aspergillus flavus*.

after-dinner mints Mint chocolates usually served with coffee after dinner.

agneau (*Fr.*) (masc.) [anyo] Lamb.

ail (*Fr.*) (masc.) [ahyuh] **Garlic**.

aile (*Fr.*) (fem.) [ahyul] Wing, e.g. *aile de poulet*, 'chicken wing'.

aïoli/ailoli (*Fr.*) (masc.) [ahyuh-oh-lee] Provençal garlic mayonnaise. (From *ail* 'garlic' and provençal dialect *oli* 'oil'.) (*Compare* **anchoiade**.)

à la (*Fr.*) [ah lah] With, in the manner of, when referring to a feminine noun, e.g. *à la carte* (*see* **carte, à la**), *à la russe* (*see* **russe, à la**) etc. (*A la* is the feminine equivalent of **au**.)

al dente (*Ital.*) [al dentay] Cooked so as to be firm not soft, when bitten, esp. referring to **pasta**. (*Dente* means 'tooth'.)

ale Top-fermented **beer**. (*See* Chapter 14.)

alfalfa sprouts [al-fal-fa] Lucerne; fine hairy sprouts used in salads and sandwiches.

al fresco (*Ital.*) Informal (meal) in the open air.

alla carbonara See **carbonara**.

allumettes (*Fr.*) (fem. pl.) [aloo-met] Small puff pastry fingers containing savoury filling served as **hors d'oeuvre**. (Literally 'matchsticks'.)

almond paste *See* **marzipan**.

à maison *See* **maison**.

amande (*Fr.*) (fem.) [amah(n)d] Almond.

amandine (*Fr.*) [amah(n)deen] With almonds.

amaretto (*Ital.*) (1) Bitter almond-flavoured **liqueur**.

(2) Small almond biscuit or **macaroon** served at the end of the meal or with desserts.

amino acid Fatty acid derived from ammonia.

amontillado (*Span.*) [amon-till-ahdoh] Medium-dry style of **sherry**.

ananas (*Fr.*) (masc. sing.) [a-nah-na] Pineapple.

anchoiade (*Fr.*) [an-cha-wahd] **Mayonnaise** flavoured with garlic and **anchovies**. From Provence (S France). (*Compare* aïoli.)

anchovy Small, very strongly flavoured fish, usually preserved in salt.

anglaise, à l' (*Fr.*) [ah longl-ayz] English-style; means different things with different items, but is most commonly applied to boiled vegetables served simply with chopped parsley and a knob of butter. (*See also* **crème anglaise**.)

Angostura Bitters Proprietary brand of aromatic bitters used as a flavouring, by the addition of a few drops only, in drinks, e.g. **Pink Gin** or **champagne cocktail**, and in cooking. (Angostura is the former name of Cuidad Bolívar, Venezuelan town where first made.)

Anna *See* **pommes Anna**.

antipasto (*Ital.*) (*pl.* antipasti) Selection of **appetisers** served prior to eating. Italian equivalent of **hors d'oeuvre**. (Literally 'before' (*anti*) 'the meal' (*pasto*).)

apéritif (*Fr.*) (masc.) [apay-rit-eef] A drink served before a meal to stimulate the appetite. (*See* Chapter 14.)

à point *See* **point, à**.

appetiser/appetizer Food served before a meal to stimulate the appetite.

apple/apricot Danish *See* **Danish pastry**.

arborio rice Italian rice used for **risotto**.

armagnac (*Fr.*) (masc.) [ar-man-yak] Fine **brandy** made in the Armagnac district, S France.

aroma (1) Fragrance, smell.

(2) Wine-tasting term. Scent of the grapes in a young wine. As wine matures, the aroma decreases and the **bouquet** increases. (*See also* **nose**.)

aromatic Fragrant, sweet-smelling.

artichoke Two quite different vegetables:

(1) **globe artichoke** Thistle-like plant with a large flower. The 'heart' and the base of the scaly leaves of the immature flower are eaten, usually with **vinaigrette**. (French *artichaut* [arti-show].)

(2) **Jerusalem artichoke** Vegetable used in soups, stews, etc. It looks like a knobbly potato or piece of fresh **ginger**. (The name has nothing to do with Jerusalem but is a corruption of *girasole*, Italian for 'sunflower'. French *topinambour* [topi(n)-am-bore].)

arugula (*Ital.*) [a-rug-u-la] **Rocket**; green salad leaf with slightly bitter flavour.

aspic Savoury jelly (natural **gelatine**) made from meat **stock**. Used for setting cold fish or meat in a mould with vegetables, and for garnishing.

assiette (*Fr.*) (fem.) [assy-ett] Plate, e.g. *une assiette de* **viandes**.

au (*Fr.*) (masc.) [oh] With, in the manner of, when referring to a masculine noun, e.g. *au beurre* (*see* **beurre, au**), *au poivre* (*see* **poivre, au**), etc. (*Au* is the masculine equivalent of à la.)

aubergine [oh-bear-zjeen] Eggplant. (*See also* **brinjal**.)

auslese (*Ger.*) [ows-layz-uh] Sweet dessert wine made from selected late-picked **riesling** grapes. (*Auslese* means 'selected'.) (*See also* **spätlese**.)

aux (*Fr.*) (plural) [oh] With. (Plural of **au** and **à la**, e.g. **feuilleté** *aux* **pommes, soupe glacée** *aux* **moules**.)

ayurvedic tea [a-yoor-vaydik] **Herbal tea** with soothing and healing qualities.

B and B Cocktail of **brandy** and **Bénédictine**.

baba (*Fr.* from *Polish*) [baa-baa] Small light yeast cake, usually containing raisins. *Babas au rhum* (**rum** babas) are soaked in **rum** syrup.

baba ghanoush (*Mid East*) Dip or **condiment** made of cooked eggplant blended with **tahini, garlic,** lemon and spices.

babychino Drink of warm milk topped with froth and a sprinkling of chocolate. (*See also* Chapter 16 (cappuccino).)

Bacardi Proprietary brand of white **rum**.

back of house The parts of the **establishment** not seen by the guests; the kitchen, **stillroom** and accounts department, as opposed to the dining-room. (*See also* **front of house**.)

bacteria Plural of bacterium. Tiny single-celled organisms often dangerous to health, but sometimes useful, e.g. in cheese-making.

bagel (*Yid.*) [bay-gul] Ring or doughnut-shaped hard bread roll.

bagna cauda (*Ital.*) [banya cowda] Hot **anchovy** dip. (Literally 'hot bath',

from Piemont in N Italy.) (*See also* **crudités**.)

baguette (*Fr.*) (fem.) [bag-ett] French bread stick.

bain-marie (*Fr.*) (masc.) [ba(n) maree] Large open dish partly filled with hot water, in which pans stand so that their contents are kept hot without overcooking. (Literally 'bath of Mary'.)

baklava/baclava (*Grk* & *Turk.*) Filo pastry cake containing almonds and spices and dipped in honey. Traditionally it is cut into triangular shapes.

ballottine (*Fr.*) (fem.) [ba-yot-teen] Meat or fish boned, stuffed, rolled and served cold with an **aspic** glaze, or in a **chaudfroid**. A ballottine may also be served hot, **glazed** in its own **reduced** juices. (*Ballot* means 'bundle'.)

Balmain bug *See* **bug**.

balsamic vinegar Fragrant, matured, sweet wine **vinegar** with intense flavour. (*See also* **malt vinegar, rice vinegar** *and* **wine vinegar**.)

balthazar Very large bottle of **sparkling wine**, the equivalent of 16 standard 750mL bottles. (Balthazar was one of the three wise men who visited the infant Jesus.) (*See also* **champagne bottle sizes**.)

bap (*UK*) Soft round white bread roll dusted with flour.

barista (1) Expert coffee-maker. (*See* Chapter 16.)
(2) (*Ital.*) Bar attendant; bar proprietor.

baron of beef Very large joint of beef, including **loin** and rump, often spit-roasted.

barquette (*Fr.*) (fem.) [bar-ket] Small boat-shaped pastry that can be filled with various ingredients. (Literally 'little boat', 'little barque'.)

basil Pungent, sweet leaves from related plants (species *Ocimum*) used as flavouring herb, esp. in Italian and SE Asian cookery. (*See also* **pesto** *and* **pistou**.)

basmati rice [bass-mutee] Fragrant long-grain rice used in Indian cuisine, esp. **pilau** and **biryani**.

batter Fluid dough of flour and water, milk or **beer**, usually with egg, used in cooking. It may be poured, e.g. to make **pancakes**, **waffles** or **Yorkshire pudding**, or it may be used as a coating, e.g. to make **fritters** or deep-fried fish.

bavarois (*Fr.*) (masc.) [bavah-rwah] Dessert of **custard** stiffened with **gelatine** mixed with whipped **cream**, shaped in a mould and served cold. (*Bavarois* means 'Bavarian', from Bavaria, state in S Germany.)

bay leaf Leaves of the bay tree used, fresh or dried, to give flavour to **casseroles**, **stocks** and **marinades**. They are also used in a traditional **bouquet garni**.

bean curd Bland curd made from **soya beans** and rich in minerals and protein. (*See also* **tofu**.)

beansprout/bean shoot Pale-coloured sprout of the mung bean (*Phaseolus aureus*), about 5cm long with a small pod at one end. It is eaten in salads, sandwiches and **stir-fries**.

béarnaise (*Fr.*) [bay-air-nayze] Sauce made of beaten egg yolks and **reduced wine vinegar** mixed with butter and served warm, usually with fish or grilled meat. (In French, *sauce béarnaise*. Béarn is a province in SW France.)

beaujolais (*Fr.*) [bow-zjol-lay] French wine from the Beaujolais region of E France, or a **generic** wine in that style. The wine is red, fruity and light, and drunk while still young. (*See below*.)

beaujolais nouveau (*Fr.*) [bow-zjol-lay noo-voh] Beaujolais wine from the latest **vintage**. Australia has borrowed from the French tradition of racing the new vintage to particular restaurants as soon as it is bottled to see who can serve it first.

béchamel (*Fr.*) (fem.) [bay-sha-mel] Sauce made by adding milk to a **roux**; it is the foundation of many other sauces. (Attributed to Louis de Béchameil, Lord Steward to Louis XVI.)

beef bourguignon *See* **boeuf bourguignon**.

beef olive *See* **paupiette**.

beef rendang *See* **rendang daging**.

beef Stroganov *See* **boeuf Stroganov**.

beef Wellington Beef fillet, liver **pâté** and mushroom **duxelles** baked in **puff pastry**. (Named in honour of the Duke of Wellington, 1769–1852, British general and Prime Minister.) (*See also* **croûte, en**.)

beer Alcoholic beverage made from **fermented** malted barley (sometimes wheat) flavoured with hops; general term for **ales**, **lagers** and **stouts**. (*See* Chapter 14.)

beet (leaves) Beetroot (leaves). The root is the key ingredient in **bortsch**. The leaves can be used as a salad vegetable.

beignet (*Fr.*) (masc.) [bay-nyay] Fritter.

Belgian endive *See* **chicory**.

Bénédictine (*Fr.*) [benny-dick-teen] Proprietary liqueur, very sweet and spicy, from Normandy (N France). (*See also* **B and B.**)

bento box Japanese lunch box containing Japanese finger food (*bento*).

bergamot Orange-scented herb (*Monarda didyma*) used in savoury and sweet dishes. It gives a distinctive flavour to **Earl Grey tea**.

beurre (*Fr.*) (masc.) [burr] Butter.

beurre, au (*Fr.*) [oh burr] Cooked in butter.

beurre blanc (*Fr.*) (masc.) [burr blo(n)] Smooth sauce made by whipping butter into a **reduced** mixture of white wine, **wine vinegar** and finely chopped onions. (*Blanc* means 'white'.)

beurre fondu (*Fr.*) (masc.) [burr fon-doo] Butter melted slowly with lemon juice, white **pepper** and salt, often served with boiled or steamed vegetables and poached fish. (*Fondu* means 'melted'.)

beurre maître d'hôtel (*Fr.*) (masc.) [burr may-truh doh-tell] Softened butter added to mushroom **duxelles**, chopped parsley and lemon juice. Usually served with fish or grilled meat. (*See* **maître d'hôtel**.)

beurre meunière (*Fr.*) (masc.) [burr murn-ee-yair] **Beurre noisette** with lemon juice added. (*See also* **meunière, à la**.)

beurre noisette (*Fr.*) (masc.) [burr nwa-zet] Sauce of butter heated until brown (nutty) and served very hot. (*Noisette* is French for 'hazelnut'.)

bianco (*Ital.*) [bee-yan-ko] Golden medium-sweet style of **vermouth**. (Literally 'white'.)

bien cuit (*Fr.*) (masc.) [bee-a(n) cwee] Over-cooked (steak). (Literally 'well cooked'.) (*See also* **bleu, point (à)** *and* **saignant**.)

bifteck (*Fr.*) (masc.) [biff-tek] Beef steak, e.g. *bifteck au* **poivre**, 'pepper steak'. (*Bifteck* is a French corruption of the English words 'beef steak'.)

billfold (*USA*) Wallet or folder used for presenting bills, change, etc. (*See* Chapter 17.)

bill of fare [fair] Menu.

bin Section of a wine cellar in which bottles of a particular wine or other product are stored.

Bircher muesli (*Swiss Ger.*) [beer-sher mew-zlee] Breakfast dish made of oats, yoghurt, honey and fruit juice. It is porridge-like in consistency but dried fruit and nuts are added to give it texture. (Developed by Dr Bircher-Benner.) (*Compare* **porridge**.)

biryani/biriani (*Ind.*) [birry-ahnee] **Pilau**, usually spiced and coloured yellow with **saffron** and garnished with hard-boiled egg.

biscotte (*Fr.*) Toasted slice of **brioche** or a rusk, served at afternoon tea.

biscotti (*Ital.*) Small flavoured biscuits, usually served with coffee.

bisque (*Fr.*) (fem.) [beesk] Thick creamy soup, usually based on **seafood**, e.g.

prawn bisque. (*See also* **chowder** *and* **potage**.)

bistro/bistrot (*Fr.*) [beess-troh] Small informal restaurant or licensed café.

bitters Flavoured alcoholic spirits of varying strengths. Different brands have very different characters and uses. (*See* **Angostura Bitters, Campari, Fernet Branca** *and* **Underberg**.)

black bean Salted **soya bean** available dried or bottled. (*See also* **black bean sauce** *below*.)

black bean sauce Sauce made from fermented **soya beans** used in Chinese cooking.

black coffee *See* Chapter 16.

black pepper *See* **pepper, peppercorn**.

black pudding (*UK*) Savoury pork sausage stuffed with oatmeal, blood, seasoned meat, etc. (*Compare* **boudin noir**.)

blanc(he) (*Fr.*) [bla(n) /blah(n)sh] White.

blanch To place briefly in boiling water and then drain to remove excess salt, bitterness, etc. before normal cooking. (Literally 'to whiten'.)

blanquette (*Fr.*) (fem.) [bla(n) -ket] Delicate but creamy stew of **veal**, lamb or chicken with vegetables, e.g. *blanquette de* **veau**. (From *blanc*, 'white'.)

blend To mix, usually in an electric blender. Common technique for mixing **cocktails**. (*See* Chapter 14.)

bleu (*Fr.*) [bluh] Extremely **rare** (steak, etc.). (Literally 'blue'.) (*See also* **bien cuit, point** (à) *and* **saignant**. *See also below*.)

bleu, au (*Fr.*) [oh bluh] Way of cooking freshly killed fish, particularly trout, so that the skin has a bluish tinge. (*Bleu* means 'blue'.)

blini (*Russ.*) [blee-nee] Small thick savoury **pancake**, traditionally served with **soured cream** to accompany **caviare**.

blintz (*Yid.*) Small savoury **pancake** like a **blini**, but often filled with cheese.

Bloody Mary Cocktail of **vodka** in **tomato** juice, flavoured with lemon juice, **Tabasco** and **Worcestershire sauce**.

blue Extremely **rare** (steak, etc.). (*See also* **bleu**.)

blue vein Blue mould in blue or green cheese. (*See* **Gippsland blue, gorgonzola, roquefort** *and* **stilton**.)

bocadillo (*Span.*) Sandwich.

bocconcini (*Ital.*) [bokon-cheenee] Small balls of fresh white **mozzarella** preserved in whey to retain moisture. Bocconcini are often served as part of the **antipasto**. (Literally 'little mouthfuls'.)

body Consistency or 'thickness' of wine (wine-tasting term).

boeuf (*Fr.*) (masc.) [burf] Beef. (*See also* **bifteck**.)

boeuf à la ficelle (*Fr.*) [burf ah lah fee-sel] Beef **fillet** tied with string to keep its shape, and cooked in **stock**. (*Ficelle* means 'string'.)

boeuf à la mode (*Fr.*) [burf ah lah mode] **Braised** beef cooked with **diced** veal, carrots and onions.

boeuf bourguignon (*Fr.*) [burf boor-gee-nyo(n)] Beef bourguignon. A casserole made with braising steak and red wine. (*Bourguignon* is French for 'Burgundian' or 'from **Burgundy**'.)

boeuf en daube (*Fr.*) [burf o(n) dobe] Beef **braised** in a red wine **stock**. (*See also* **daube, en**.)

boeuf Stroganov/Stroganoff Strips of beef in a creamy sauce garnished with mushrooms. (The Stroganovs were a wealthy merchant family ennobled by Peter the Great of Russia.)

bok choy (*Chin.*) White cabbage with fleshy white stems and light green leaves.

bolognais(e) (*Fr.*) [bol-on-ayz] With a thick meat-and-**tomato** sauce; properly *à la bolognaise* in French, meaning 'in the style of Bologna' (city in NE Italy).

bolognese (*Ital.*) [bolon-ayz-ay] Italian for **bolognaise**. (In Italy a bolognaise sauce is usually called *ragu*.)

bombe (*Fr.*) (fem.) [bombuh] Frozen dessert (*bombe* **glacée**) made in a spherical mould (or one with a rounded top or, loosely, any mould) lined with ice cream and various fillings. (*Bombe* is French for 'bomb', or 'spherical container'.)

bon appétit! (*Fr.*) [bo(n) apay-tee] Enjoy your meal! (Literally 'good appetite'.)

bonne femme (*Fr.*) [bon fam] Simply cooked; home-style, with potato, e.g. **potage** *bonne femme*. (Literally 'good woman'.)

Bonsoy Proprietary name for a brand of Japanese soy drink; milk alternative.

bordelaise, à la (*Fr.*) [ah lah bord-eh-layz] In the style of Bordeaux (town in SW France); cooked with **shallots** and wine, e.g. **poulet** *sauté à la bordelaise*.

börek (*Turk.*) Pasty made with **filo** or layered pastry. Although more commonly savoury, sweet versions are available. Börek was originally a Turkish pastry, but there are many variations from other countries, esp. the Balkans and N Africa.

bortsch/borsh (*Russ.*) [borsch] Beetroot soup served with **soured cream**.

botrytis Fungus or mould (*Botrytis cinera*), sometimes called **noble rot**, which shrivels late-picked grapes and makes them intensely sweet. Botrytis-affected grapes are used to make many excellent sweet **dessert wines**. (*See* Chapter 14.)

bottomless cup System used for **coffee**, etc. whereby customers may have their cups refilled as often as they like at no additional charge.

botulism Food poisoning found principally in canned food and sausages, caused by the bacterium *Clostridium botulinum*. (*Botulus* is Latin for 'sausage'.)

bouchée (*Fr.*) (fem.) [boo-shay] Small **puff pastry** appetiser containing a savoury filling. (Literally a 'mouthful'; *bouche* means 'mouth'.)

boudin (*Cajun*) [booda(n)] Sausage made of pork liver, cooked pork, rice and seasonings.

boudin noir (*Fr.*) [booda(n) nwa] **Black pudding**; sausage made from pig's blood, **cream**, fat and seasonings. (*Noir* means 'black'.)

bouillabaisse (*Fr.*) (fem.) [boo-ya-bayss] Rich **provençal** stew of fish, mussels, etc. simmered with **herbs**. (*See also* **gumbo** *and* **pochouse**.)

bouillon (*Fr.*) (masc.) [boo-y-o(n)] Plain unclarified meat or vegetable **broth** used as **stock** in cooking. (*See also* **consommé, julienne, potage** *and* **soupe**.)

bouquet (*Fr.*) (masc.) [boo-kay] Wine-tasting term. The smell of a maturing wine (as opposed to the **aroma** of the grapes). (*See also* **nose**.)

bouquet garni (*Fr.*) (masc.) [boo-kay gar-nee] Bunch or faggot of herbs, traditionally consisting of **thyme, marjoram**, parsley and **bay leaves**, used for flavouring sauces, **stock**, etc.

bourbon [burbun] American **whiskey** made principally from fermented maize grain. (Originally from Bourbon County, Kentucky.)

bourguignon *See* **boeuf bourguignon**.

braise To stew meat etc. slowly in very little liquid in a closed pan.

brandy Distilled **spirit** made from fermented grapes. (*See* **armagnac, cognac** *and* Chapter 14; *see also* **calvados** *and* **cherry brandy**.)

brandy snap Golden-brown lacy biscuit flavoured with **ginger** and rolled into a cylindrical shape; they are sometimes made in the form of a basket in which to serve ice cream desserts.

brasserie Café selling alcoholic beverages, esp. beer, as well as coffee and food. (The word originally meant a brewery. *Brasser* is French for 'to brew'.)

bratwurst (*Ger.*) [brat-voorst] Highly seasoned sausage made from pork and **veal**. (Literally 'frying sausage'.)

brawn Pieces of meat, esp. calf's or pig's head, cooked and set in **aspic**, shaped into a loaf and served cold. (*See also* **galantine**.)

bread and butter pudding (*UK*) Sweet **pudding** of bread layered with sultanas, candied (mixed) peel and sugar, cooked in egg **custard**.

breathe Wine-tasting term. To allow a wine to 'breathe' is to allow it to come into contact with air by removing the cork from the bottle some time before the wine is drunk in order to enhance the **bouquet**.

bresaola (*Ital.*) [brez-ay-ohla] Air-dried beef, usually sliced wafer-thin and served with antipasto.

brie (*Fr.*) [bree] Soft creamy cow's-milk cheese with a soft edible crust. (Brie is a small town near Paris.) (*See also* **camembert, chèvre** *and* **King Island**.)

brigade The staff in the dining-room or kitchen as an organised team. (*See* **rang** *and* Chapter 1.)

brigade de cuisine (Fr.) [bri-gaad duh kweezeen] The kitchen **brigade**, consisting of the chef and all his or her assistants.

brine Solution of salt and water usually used as a preservative. (*See also* **cure, pickle** *and* **smoke**.)

brinjal Word of Portuguese origin used in India and Africa for the **aubergine** or eggplant.

brioche (*Fr.*) (fem.) [bree-osh] Light soft roll made from yeast dough with eggs and butter.

broccolini Green vegetable with small broccoli heads on long (20cm), thin, branching stalks.

broche, à la (*Fr.*) [ah lah brosh] Cooked on a spit or skewer.

brochette (*Fr.*) (fem.) [bro-shett] Skewer on which pieces of meat, etc. are cooked, e.g. **kebabs**.

brodo (*Ital.*) [brohdoh] **Clear soup; stock.** (*See also* **minestra** *and* **zuppa**.)

broil To grill.

broth Meat, fish or vegetable **stock**.

brûlé(e) (*Fr.*) [broo-lay] Literally 'burnt'. (*See* **crème brûlée**.)

brunch A substantial late-morning meal, replacing breakfast and lunch. (*Compare* **yum cha**.)

brunoise (*Fr.*) (fem.) [broon-warze] **Diced** vegetables, often **braised** in butter, used as a **garnish** for soups, sauces, etc. (*Brun* means 'brown'.)

bruschetta (*Ital.*) [broo-sketta] Baked or toasted slices of bread, oiled and sprinkled with herbs and served as an **appetiser**. They are often served with a savoury topping. (*See also* **canapé** *and* **crostino**.)

brush down To remove crumbs, etc. from tables or tablecloths prior to service of the next course; also called **crumbing down**. (*See* Chapter 11.)

brut (*Fr.*) [broot] Very dry (**sparkling wine**).

bubble and squeak (*UK*) Mashed potatoes mixed with cooked cabbage and fried. (*Compare* **champ** *and* **colcannon**; *see also* **réchauffé**.)

bubble tea Icy flavoured drink containing balls of **tapioca**.

buckwheat Not strictly speaking a **cereal**, but the round plump seeds of an annual plant (*Fagopyrum esculentum*) used in a similar way. It has a strong, distinctive taste and is much used in Chinese, E European and Jewish cuisine. Buckwheat flour is used to make **blini** and **piroshki**.

buffet [booff-ay] (1) Meal consisting of a number of dishes set out so that guests can select what they want for themselves. (*See* Chapter 10 *and* **smorgasbord**.)
(2) Room or counter where snacks or light meals may be bought.

buffet froid (*Fr.*) (masc.) [boof-ay fwah] Cold meats or shellfish; one of the later courses of the **classic menu**, following the salads and preceding the sweets (**entremets**).

buffet, turkey *See* **turkey buffet**.

bug Shellfish with flesh similar to **rock lobster**, but without long legs and feelers. There are two common varieties: Balmain bug and Moreton Bay bug.

build Cocktail-mixing term. To add ingredients one to the other in the glass in which they will be served. (*See* Chapter 14.)

bullboar (*Austral.*) Thick, spicy pork and beef sausage made with red wine. A speciality of Central Victoria.

burghul (*Mid East* & *Grk*) Style of **cracked wheat**, where the grains are hulled (removed from their husks), steamed and cracked. (*See also* **kibbeh**.)

burgundy Smooth, soft red wine from Burgundy (province in W France) or similar in style. There are also white burgundies, always so-called. (*See also* **boeuf bourguignon, chablis, pinot noir** *and* Chapter 14.)

burrito (*Mex.*) [bu-reetoh] **Tortilla**; flat bread wrapped around a filling, usually served in a cheese sauce.

bus (*USA*) To perform general clearing duties in a food or beverage service area.

busboy/busgirl/busperson Person who clears crockery, glassware, etc. from a food and beverage service area. (*Colloq.* 'bussy'.)

bush tomato/bush raisin (*Austral.*) Yellow fruit of *Solanum centrale*, light brown once dried. Bush-tomato **chutney** is available commercially.

butterfly (~ied) Cut of meat or fish, esp. prawns, split so that one side remains intact while it is opened flat on the barbecue or grill.

byessar (*Grk*) **Purée** of fava (broad) beans, usually served as part of a **mezze** plate.

BYO Bring Your Own (alcohol); restaurant to which you can bring your own alcohol.

cabernet sauvignon [cab-air-nay so-vee-nyon] Grape variety used to make red wines, esp. **claret**. Called simply 'cabernet' when blended with other varieties, e.g. cabernet-**malbec**.

cacciatore/cacciatora (*Ital.*) [cat-cha-toray] Cooked with **tomatoes**, **mushrooms**, herbs and, usually, wine, e.g. **pollo** *alla cacciatora*, chicken cacciatora. (*Cacciatore* means 'hunter'.) (*See also* **chasseur**.)

Caesar salad Salad consisting of **cos** lettuce, dressing, almost raw eggs, **parmesan** cheese, diced crisp bacon, **anchovies** and **croûtons**, often served as a substantial first course. (*See also*

Chapter 12.) (Devised by Caesar Cardini, chef in Tijuana, Mexico, *c.* 1925.)

café au lait (*Fr.*) (masc.) [caffay oh lay] White **coffee**; coffee with hot milk or **cream** added. (*See* Chapter 14.)

café complet (*Fr.*) [caffay complay] A pot of coffee and milk served with **croissants**, butter and **preserves**.

caffeine [caf-feen] Mild stimulant found in **coffee** and **tea**.

caffè frappé (*Ital.*) [caffay frappay] Milkshake; ice-cold drink made with milk, espresso coffee and coffee syrup blended with ice to make a long drink with a grainy texture. (*Compare* **affogato**, **granita** and Chapter 16 (iced coffee).)

caffè latte (*Ital.*) [caffay lahtay] *See* Chapter 16.

cafeteria A self-service restaurant.

Cajun/Cajan [kay-jun] **Cuisine** developed by **creole** French-speakers in Louisiana (state in S USA). The key ingredients are **capsicum**, onion and celery with plenty of **pepper**. (*See also* **gumbo**.)

calabrese (*Ital.*) (1) Green sprouting broccoli. (2) Hot spicy **salami**. (Calabrese means 'Calabrian'— Calabria, region in S Italy.)

calamari (*Ital.*) (*pl.*) [kal-a-maar-ee] Squid.

call away/call up To request the kitchen to plate up the next **course**.

Calorie Obsolescent measure of the energy in food. Short for kilocalorie. 1 Calorie = 4.2 **kilojoules**.

calvados Apple **brandy** made in Normandy (province in N France famous for apples). (*See* **normande**.)

calzone (*Ital.*) [kal-zohny] Savoury pasty made with **pizza** dough.

camembert (*Fr.*) [cam-embear] Soft cheese made from cow's milk, with a downy skin. (Camembert is a village in Normandy, N France.) (*See also* **brie, chèvre** *and* **King Island**.)

Campari Brand of Italian **bitters**. Bright red, it is often served as an **apéritif** mixed with **soda**.

canapé (*Fr.*) (masc.) [can-a-pay] Small piece of bread (usually toasted) or biscuit, garnished with **caviare**, cheese, **pâté, smoked** salmon or similar. Cold canapés are usually served as **appetisers** with drinks before a meal; hot canapés are sometimes served as **entrées** in the meal itself. (*See also* **bruschetta, crostino** *and* **Melba toast**.)

canard (*Fr.*) (masc.) [can-ar] Duck. (*See also* **caneton**.)

cane spirit Clear spirit distilled from sugar, often used in **cocktails**.

caneton (*Fr.*) (masc.) [can-uh-to(n)] Duckling. (*See also* **canard**.)

cannelloni (*Ital.*) (*pl.*) Large tubes of **pasta** filled with cheese, meat, fish, etc. and baked in a sauce.

cantaloup/cantaloupe Rockmelon.

canteen (1) Type of restaurant in a factory, school, etc.
(2) Box containing a complete set of cutlery for several place settings.

Cantonese One of the five main styles of Chinese cuisine; *Cantonese* cuisine is the style usually encountered in Westernised restaurants. (*See also* **Fukien, Honan, Peking** *and* **Szechwan**.) (Canton is the old name

for a city in S China near Hong Kong, now called Guangzhou.)

caper Tiny green **pickled** bud of the caper bush (*Capparis spinosa*). Capers are preserved in either salt or brine and are used as a **garnish** or seasoning. (*Compare* **caper berry**.)

caper berry Pickled, olive-sized fruit of the caper bush (*Capparis spinosa*) preserved in brine. Caper berries are served as an appetising snack or as an **hors d'oeuvre**. (*Compare* **caper**.)

caponata (*Ital.*) [capoh-nahtah] Dish of eggplant, tomatoes and celery flavoured with **capers, anchovies** and **olives** fried together in olive oil but served cold, usually as an **hors d'oeuvre**.

cappuccino (*Ital.*) [cap-poo-chee-no] (*See* Chapter 16.)

capretto (*Ital.*) [kap-rettoh] Milk-fed kid (young goat), usually up to 5kg in weight. (*Compare* **chevron**.)

capricciosa (*Ital.*) [caprit-chosa] Sauce containing **tomato**, cheese, ham, mushrooms, **olives** and, usually, **anchovies** commonly served as **pizza** topping. (*Capriccioso* means 'naughty' or 'capricious'.)

capsicum Family of plants otherwise called **peppers**, including **chillies** and **sweet** (or bell) **peppers**. The word, used alone, usually refers to the latter.

carafe (*Fr.*) (fem.) [ca-raaf] Glass bottle or spoutless jug used for water or wine.

caramel Melted sugar heated slowly until it is rich and brown.

caramelise (1) To turn sugar into **caramel** by heating.
(2) To pour melted sugar over food to brown it.

carbohydrate Essential energy-giving **nutrient**; an organic compound of carbon, hydrogen and oxygen. (Starch, sugar and cellulose are groups of carbohydrates.)

carbonara (*Ital.*) Sauce made from **pancetta** or bacon, cheese and raw egg served on **pasta**, esp. **spaghetti** (spaghetti alla carbonara). The hot pasta cooks the egg and melts the cheese.

carbonated wine Wine made effervescent (bubbly) by pumping carbon dioxide into it. (*See* Chapter 14 *and* **sparkling wine**.)

Caro Trade name of a naturally **caffeine**-free instant **cereal** and **chicory** beverage used as a substitute for instant **coffee**.

carob Pod of *Ceratonia siliqua* (locust bean) used when ground as a sugar or chocolate substitute.

carpaccio (*Ital.*) [car-patchy-oh] Thin slice of raw beef or fish, e.g. tuna, served cold.

carré (*Fr.*) (masc.) [caray] Cut of meat, e.g. *carré d'*agneau (**rack** of lamb), *carré de* **veau** (neck of **veal**).

carte, à la (*Fr.*) [ah lah kart] Type of **menu** offering a choice of items that are individually priced and cooked to order. (*See* Chapter 2.)

carte du jour (*Fr.*) (fem.) [cart doo joor] **Menu** of the day. (*See* Chapter 2.)

cassata (*Ital.*) [kass-ahta] Layers of different ice creams, one of which is flavoured with **liqueur** and contains **glacé** fruits. (Literally 'little case' because of its traditional brick shape.)

casserole (1) Pan with a lid for cooking stews. (2) Stew cooked in a casserole.

cassis (*Fr.*) (masc.) [cass-eece] (1) Blackcurrant. (2) Blackcurrant **liqueur**.

cassoulet (*Fr.*) (masc.) [cass-oo-lay] **Râgout** of **haricot** beans and meat with a **gratin** topping.

catsup *See* **ketchup**.

caviare/caviar Salted sturgeon's eggs. Other fish **roe**, esp. **lump-fish**, is often used as a substitute.

cayenne pepper Hot spice made from dried, ground red **chillies**, sometimes placed on the table in a small **cruet**. Also called 'red pepper' or 'chilli powder'. (Originally from Cayenne, a city in French Guyana.)

celeriac [sell-air-eeyac] Root vegetable, about the size of a large potato, with rough pale brown skin. It is usually roasted or **mashed** and is often served *en rémoulade* (blanched and in a **rémoulade** sauce).

cep [sep] Mushroom with thick stalk (*Boletus edulis*). (French *cèpe* [saip]; Italian *porcino* [pore-cheeno].) (*See also* **morel**, **oyster mushroom**, **pine mushroom**, **shitake** *and* **truffle**.)

cereal [seerial] (1) Any grain used as food; corn: wheat, barley, maize, oats, rye, etc.
(2) Packaged precooked breakfast food made from such grain. (Ceres was the Roman goddess of corn crops.)

cerise (*Fr.*) (fem.) [ser-reece] Cherry.

cervelles (*Fr.*) (fem. pl.) [sur-vel] Brains.

cevapcici (*Serb.*) [sevap-chee-chee] Skinless spicy sausage or meatball made of several different meats minced with **capsicum** and **garlic**.

chablis [shablee] Dry white **table wine** made in the northern **Burgundy**

district of France, or a **generic** wine in that style.

chafing dish [chay-fing] (1) Portable pan with a source of heat used for cooking at table, or on a **guéridon**. (2) Hotplate for keeping dishes warm; a **réchaud**. (*See* Chapter 3.)

cha gio (*Viet.*) [chah zo] **Spring rolls**; small **pancake** parcel filled with minced crab, pork and mushroom, deep-fried and wrapped in a lettuce leaf with **Vietnamese mint**. They are served with **nuoc cham** as a dipping sauce.

chai (*Ind.*) [chy—rhymes with 'eye'] Sweet green **tea** made with milk and added spices, esp. cinnamon, cloves, cardamom and **saffron**. Also known as *masala* chai.

champ (*Irish*) Potato mashed with cream and spring onions or **chives**. (*Compare* **bubble and squeak** *and* **colcannon**.)

champagne Sparkling wine from the Champagne region (N France). (*See* Chapter 14.)

champagne bottle sizes Champagne and other sparkling wines are often available in extra large bottles, with traditional names: **magnum** (2 bottles), **jeroboam** (4 bottles), **rehoboam** (6 bottles), **methuselah** (8 bottles), **salmanazar** (12 bottles), **balthazar** (16 bottles) and **nebuchadnezzar** (20 bottles). (The 'bottles' are multiples of the standard 750mL bottle.)

champagne cocktail Cocktail of **sparkling wine** with a small measure of **brandy**, a sugar cube and a few drops of **Angostura Bitters**.

champignon (*Fr.*) (masc.) [shom-pee-nyo(n)] Mushroom.

chantilly *See* **crème chantilly**.

chapati/chupatty (*Ind.*) Thin unleavened bread; **roti**.

charcoal-grill To grill food over a 'natural' heat source of burning charcoal. (*See also* **char-grill**.)

charcuterie (*Fr.*) (fem.) [shar-coot-er-ee] (1) Pork butchery. (2) Cooked meats, e.g. **terrines**, **smoked** ham, **salamis**, etc., made primarily from pork. (*See also* **smallgoods**.)

charcutière, à la (*Fr.*) [ah lah shar-coot-ee-air] Served with a **demi-glace** containing wine and thin strips of **dill pickle**.

chard, rainbow *See* **silverbeet**.

chardonnay [shar-don-ay] Grape variety used to make white wines, e.g. white **burgundy**.

char-grill To cook food on a grill, which has coke or coals, over an 'artificial' electric or gas heat source. (*See also* **charcoal-grill**.)

charlotte (*Fr.*) (fem.) [sharl-ot] Hot baked dessert of **puréed** fruit cased in or layered with bread or sponge cake, cooked in a mould. (Named in honour of Queen Charlotte, consort of George III.)

charlotte russe [sharl-ot rooss] Cold dessert of **custard** prepared in a **charlotte** mould lined with jelly and sponge fingers. (*Russe* means 'Russian'.)

Chartreuse Old French **proprietary liqueur** with very complex flavours.

There are two styles: green and yellow. The green is stronger and less sweet.

chasseur (*Fr.*) (masc.) [shass-ur] Cooked with white wine, **shallots**, mushrooms and tomatoes, e.g. **tournedos** *chasseur*. (*Chasseur* means 'hunter'.)

châteaubriand (*Fr.*) (masc.) [shat-oh-bree-ond] Tender thick **fillet** steak, traditionally served for two and cut at the table, usually served with a **béarnaise** sauce. (Devised by the chef to the Vicomte de Châteaubriand, 1768–1848, French diplomat, writer and gourmet.)

chaud (*Fr.*) [show] Hot.

chaudfroid (*Fr.*) (masc.) [show-frwah] Cold dish of fish or meat served in a coating of **gelatine** (aspic); cooked hot and served cold. (*Chaud* is French for 'hot' and *froid* is French for 'cold', so literally 'hot-cold'.)

chawarma/shwarma (*Leb.*) [sha-warma] *See* **döner kebab**.

cheddar The most popular kind of tasty cheese, originally made in the village of Cheddar in Somerset, England.

chef (*Fr.*) (masc.) [shef] Senior or head cook. Short for *chef de* **cuisine**. (*Chef* is French for 'chief' or 'head'.)

chef de partie (*Fr.*) [shef duh partee] Qualified cook or chef.

chef de rang (*Fr.*) (masc.) [shef duh rung] Traditional term for a senior waiter; the waiter immediately under the station head waiter in an **establishment** with a full service **brigade**. (*See* **rang** *and* Chapter 1.)

chermoula (*Moroccan*) [chair-moolah] Traditional **marinade** for fish made with fresh **coriander**, spices, onion, **garlic**, lemon and oil. In modern cuisine it is often used to flavour chicken dishes or served on its own as a **condiment**.

cherry brandy Cherry-flavoured **liqueur**, usually made by **macerating** cherries in neutral spirit. (*See also* **kirsch**.)

cherry tomato *See* **tomato**.

chervil Small plant with parsley-like leaves used as a herb for seasoning, esp. in French cuisine; one of the ingredients of **fines herbes**.

chèvre (*Fr.*) (fem.) [shevruh] (1) She-goat. (2) Goat's cheese.

chevreuil, en (*Fr.*) [ahn shev-ruyee] Cooked to taste like **venison**; **marinated** meat served with *sauce* **poivrade** or *sauce* **venaison**. (*Un chevreuil* is a roe-deer.) (*See also* **salmis**.)

chevron (*Fr.*) [shev-roh(n)] Young goat, usually up to 16kg in weight. (*Compare* **capretto**.)

chiboust *See* **crème chiboust**.

chick-pea Seed of the **legume** *Cicer arietinum*. It is a staple of Middle Eastern cooking, and is used in dishes such as **hummus** and **falafel**.

chicory (1) Belgian or white **endive** (*Cichorium intybus*), also called **witloof**; vegetable with tightly wrapped white leaves that may be thinly sliced and served raw in salad, or cooked by steaming or boiling. Its roots are ground and mixed with **coffee** and it is used in coffee substitutes such as **Caro**. (2) (*UK*) Curly **endive** (*Cichorium endivia*); sharply flavoured spinach-like vegetable. (Italian *cicoria;* Greek *rathikia.*)

chiffonnade (*Fr.*) (fem.) [shiffon-ard] Leaves of salad vegetables (lettuce, **chicory**, etc.) cut or shredded into thin strips, used as a **garnish**.

chilli Pod and seeds of varieties of the **capsicum** plant used as a very hot spice, particularly in SE Asian, Indian and Mexican food. When dried and powdered, red chillies become **cayenne pepper**. (*See also* **pepper** *and* **sweet pepper**.)

chilli con carne (*Mex.*) [chilly con car-nay] Stew of **chillies, minced meat** and beans. (*Chile con carne* is Spanish for 'chilli with meat'.)

China tea Green tea, often served without milk. (*See also* **jasmine tea**.)

Chinese parsley *See* **coriander**.

chipolata [chipoh-lahta] Very small sausage.

chipotle (*Mex.*) [chee-poh-tul] Dried, smoked **jalapeno** chilli. Chipotles are wrinkled and reddish brown once preserved in this way.

chives Grass-like **herb** related to, and tasting of, onion and **garlic**, but without a distinct 'bulb'; it is one of the ingredients of **fines herbes**.

chlorophyll Substance that makes plants green. (Literally 'green leaf' in Greek.)

choi *See* **choy**.

cholesterol [kolesterol] Fatty substance found only in animals, high levels of which are associated with heart attacks.

chop suey (*Chin. & USA*) [chop sooee] Shredded meat cooked with vegetables.

chorizo (*Span. & Mex.*) [koreet-soh] Spicy pork sausage, usually eaten cooked, esp. with **paella**.

chou (*Fr.*) (masc.) [shoo] Cabbage. (*See also* **choy**.)

choux (*Fr.*) (masc. pl.) [shoo] Small light buns made of **choux pastry**. Choux are served either cold as a **sweet** with a **cream** filling, or with a savoury stuffing as **appetisers**.

choux pastry [shoo] Very light pastry used to make **choux, éclairs, profiteroles**, etc.; **pâte à choux**.

chowder (*USA*) Stew or thick soup, usually milk-based and made with **seafood**, e.g. clam chowder. (*See also* **bisque**.)

chow mein (*Chin. & USA*) [chow meen] Fried **noodles** with green vegetables and, usually, meat.

choy (*Chin.*) Cabbage. (*See also* **bok choy** *and* **choy sum** *below.*)

choy sum (*Chin.*) Pretty variety of cabbage with a long stem and yellow flowers. The stems are particularly tasty, with a mildly bitter flavour.

Christmas pudding (*UK*) Rich steamed pudding containing dried, fresh and **glacé** fruits, spices, nuts, **suet** and **brandy**. It is made months in advance and reheated on Christmas Day.

chutney (*Anglo-Ind.*) Thick mixture of fruits and vegetables preserved in **vinegar**, sugar and spices, served as a **condiment**.

ciabatta (*Ital.*) [chee-a-bahta] Bread with chewy texture that doesn't rise as it is baked. (Literally 'slipper', because it is said to resemble one.)

citronelle *See* **lemon grass**.

citrus Fruit from any citrus tree—lemon, lime, orange, grapefruit, cumquat, etc.

clafoutis (*Fr.*) [claff-ootee] Baked **pudding** of **batter** spread thickly on a base of fruit, esp. black **morello** cherries, and served warm.

claret Red wine from the Bordeaux region of SW France. Clarets are medium-bodied and have a distinctively astringent aftertaste because of their high **tannin** content. (The English word 'claret' comes from the French word *clairet*, meaning 'clear, bright' (wine).) (*See also* **cabernet sauvignon**.)

clarify To make clear; to remove the impurities from **stocks**, etc. so that they are no longer cloudy. **Stocks** may be clarified by boiling egg whites in them, which coagulate and trap the particles to be removed. (*See also* **ghee**.)

classic menu The full traditional banquet menu as it developed in France in the 19th century, with courses offered as follows: (1) **hors d'oeuvres** (2) soups (**potages**) (3) egg dishes (**oeufs**) (4) **pasta** and rice dishes (**farinaceous** dishes) (5) fish (**poissons**) (6) **entrées** (meaning small, preliminary meat dishes, the first of the meat courses, *see* Chapter 2) (7) **sorbets** (8) **relevés** (9) roasts (**rôtis**) (10) vegetables (**légumes**) (11) salads (12) cold buffet (**buffet froid**) (13) sweets (**entremets**) (14) **savouries** (15) cheeses (**fromages**) (16) desserts (in the sense of fresh fruit and nuts, *see* Chapter 2) (17) beverages (coffee, tea, etc.—not, strictly speaking, a course).

clear soup Consommé. (*See also* **brodo, julienne, marmite petite, steamboat** *and* **tom yam**.)

cloche Dish cover, usually metal, often hemispherical with a handle at top, serving to keep food warm. A cheese cloche (to protect from flies, etc.) is usually made of glass or transparent plastic. (*See* Chapter 3.)

clotted cream *See* **cream, clotted**.

club sandwich Sandwich composed of several layers spread with different fillings. They can be uncooked or toasted.

cocktail Mixed drink, almost always alcoholic, often with a **spirit** base. (*See* Chapter 14.)

coconut Edible flesh of the fruit of the coconut palm. It can be **desiccated** (dried), shredded and used in **curries**, cakes, **puddings** and confectionery.

coconut cream Thick mixture of processed **coconut** flesh and **coconut milk** used in **curries** and Asian dishes.

coconut milk (1) Liquid from inside the **coconut**.
(2) Processed milky mixture of **coconut** flesh with **coconut milk** used as a drink or in Asian cooking.

cocotte (*Fr.*) (fem.) [cock-ot] Small round or oval ovenproof dish in which food is cooked and served. (*See below*.)

cocotte, en (*Fr.*) [ahn cock-ot] Cooked and served in a small **cocotte** dish, e.g. **oeufs en cocottes**.

coeur à la crème (*Fr.*) [cur ah lah crem] Dessert of **fromage blanc**, drained and set in a perforated heart-shaped mould and turned out to serve with fresh fruit or a **coulis**.

coffee Drink made from the ground beans of the coffee (*Coffea*) plant, usually served after a meal. There are many ways of serving it. (*See* Chapters 14, 15 and 16; s*ee also* **affogato, caffè frappé, café au lait, Cona, corretto, decaffeinated, doppio, granita, Greek, liqueur, lungo, mochaccino, tartuffo** *and* **Turkish coffee.**)

cognac (*Fr.*) (masc.) [con-yak] Fine **brandy** from the Cognac region of SW France.

Cointreau (*Fr.*) [kwa(n)-troh] A popular **proprietary** French **triple sec curaçao liqueur.**

Colada One of a group of **blended** long-drink **cocktails**, all which include white **rum, coconut cream**, pineapple juice and cream.

colcannon (*Irish*) Mashed potatoes flavoured with onions or leeks and mixed with cooked cabbage. (*Compare* **champ** *and* **bubble and squeak**.)

coleslaw Finely shredded white cabbage with **mayonnaise.**

collation A light informal meal, often served outside normal service hours.

com chien (*Viet.*) [kom chin] Fried rice.

commis (*Fr.*) (masc.) [commie] Assistant or trainee. Short for commis cook or *commis de cuisine*.

commis de rang (*Fr.*) (masc.) [commie duh rung] Assistant or trainee waiter, below the **chef de rang** in the traditional service **brigade**.

commis waiter [commie] Ordinary assistant waiter or trainee waiter. (*See* **commis** *and* **commis de rang**.)

compote (*Fr.*) (fem.) [kompot] Fruit cooked in syrup; stewed fruit.

com tam (*Viet.*) [kom tum] 'Broken' rice topped with pork. (*Com* is 'rice', *tam* 'crushed'.)

Cona coffee Method of making **coffee** by vacuum **infusion**, similar in principle to the percolator method. The equipment consists of two bowls that are placed one above the other. Water is boiled so that it rises from the lower bowl through a tube into the upper bowl where it passes through ground coffee and a filter before returning to the lower bowl from which it is served. Cona is a trade name. (*See* Chapter 3.)

concassé (*Fr.*) [kon-kassay] Crumbled; crushed.

condé (*Fr.*) [conday] Various dishes dedicated to the Prince de Condé, great French general (1621–86) or his descendants. A condé is usually a cold dessert of poached fruit, esp. apricots, arranged in and around a ring of sweet, creamy rice and served cold.

condiment Seasoning, e.g. **mustard, soya sauce** or **pickle**, used to give relish to food, usually added to the food after it has been served at table.

confit (*Fr.*) (masc.) [con-fee] (1) Piece of almost boneless meat, usually duck, goose or **game**, cooked in its own fat and juice and sealed in a pot, immersed in the same fat, for preservation, e.g. *confit d'*oie. (2) Sweet stew of vegetables, esp. onions, served as a sauce or **condiment**, e.g. red onion confit.

congee (*Chin.*) [kon-jee] Thick rice soup topped with a savoury flavouring, e.g.

shitake mushrooms, tamari, coriander, spring onion, black beans or fried garlic.

consommé (*Fr.*) (masc.) [con-som-ay] Thin soup made from clarified **stock**, usually beef stock. It is normally served hot but may be chilled, in which case it will be jelly-like but not solid. (*See also* **bisque, bouillon, garbure, julienne, marmite petite, pistou, potage, soupe** *and* **vichyssoise.**)

contaminate (verb) hence **contamination** (noun) (1) To pollute or infect. (2) To blend different mixtures thus spoiling both. This often occurs when one utensil is reused in the preparation of different dishes.

continental breakfast Light breakfast of fruit juice, **croissants** or toast with butter, **preserves** and coffee.

coq au vin (*Fr.*) [cock oh va(n)] Chicken **casseroled** with bacon, onion, **garlic** and **mushrooms** in **brandy** and red wine. (Literally 'cock in wine'.)

coquille (*Fr.*) (fem.) [coh-kee] Shell (of fish, etc.).

coquilles Saint (St) Jacques (*Fr.*) [coh-kee san zjak] **Scallops** served cooked in a creamy sauce and presented in a scallop shell with mashed potato **piped** around the edge. (The scallop shell was the symbol of the pilgrims who visited the tomb of St Jacques—St James—of Compostella, N Spain.)

cordial (1) Any stimulating beverage. (2) Concentrated fruit squash to which water or soda is added, e.g. lemon cordial. (3) (*USA*) **Liqueur.**

coriander Fragrant herb or spice with a bitter flavour, also called **Chinese parsley**. Both the leaves and seeds are commonly used in Asian **cuisine.**

corkage Charge made by an **establishment** for opening bottles brought by customers.

corked/corky wine Wine that has been spoiled because the bottle has a faulty and mouldy cork. Seriously corked wine becomes bitter and foul-smelling, but mild corking may be difficult to detect. The word has nothing to do with pieces of cork that may fall into the wine when the bottle is opened.

corn (on the cob)/sweet corn/Indian corn/maize Cluster of small yellow grains on a central core or cob covered with fine silky threads and a green sheath. If served whole a **fingerbowl** is required. Miniature baby corn is used whole esp. in Chinese dishes. (In French, *maïs.*) (*See also* **polenta.**)

corn chip Thin crisp wafer made from maize (**corn**) flour and served with **nachos** or eaten as a snack.

cornichon (*Fr.*) [korn-ee-sho(n)] Pickled gherkin (small cucumber).

corretto (*Ital.*) Small cup or **demitasse** of espresso coffee with a shot of **spirit** (usually **brandy**) added. (Literally 'correct' or 'proper'.) (*Compare* **liqueur coffee.**)

cos [koz] Lettuce with crisp, long leaves.

côte (*Fr.*) (fem.) [coat] Rib, e.g. *côtes de veau.*

côtelette (*Fr.*) (fem.) [coat-let] Cutlet.

coulibiac (*Fr.*) [cooly-bee-ahk] A French adaptation of a Russian fish pie, traditionally layered salmon, rice and chopped hard-boiled eggs rolled in pastry.

coulis (*Fr.*) (masc.) [coo-lee] Thin **purée** of fruit, seasoned cooked vegetables, or fish.

coupe (*Fr.*) (fem.) [coop] (1) Stemmed bowl, usually made of glass. (*See* Chapter 3.)
(2) Dessert served in such a bowl (*coupe* d'ananas, *coupes* glacées).

courgette (*Fr.*) (fem.) [koor-zjet] **Zucchini**.

course Stages of a meal, e.g. **entrée**, main course, **dessert**. (*See* Chapter 2 *and* **classic menu**.)

court-bouillon (*Fr.*) (masc.) [coar bwee-yo(n)] **Broth** of white wine, herbs and vegetables in which fish is simmered; a seasoned fish **stock**.

couscous (*Arab.*) [koos koos] N African dish of steamed **semolina** pellets, usually served with meat or vegetables.

couverture Fine cooking chocolate with high cocoa butter content. (*See also* **ganache** *and* **Valrhona**.)

cover (1) Place setting for one guest.
(2) The number of guests at a **function** or **establishment**. (*See* Chapter 4.)

cover charge A charge added to the bill for rolls, butter, laundry, etc.

cracked wheat Whole wheat grains cracked and broken open by crushing to produce a coarse meal. (*See also* **burghul** *and* **kibbeh**.)

crackling Crisp baked skin or rind of pork, usually served sliced.

crayfish/crawfish/(informal) **cray** Southern **rock lobster** (*Jasus novae-hollandiae*); large spiny salt-water **crustacean** without large claws, in Australia commonly (and somewhat misleadingly) referred to simply as 'lobster'. The French for the salt-water crayfish is **langouste**. There are other varieties of freshwater crayfish (French **écrivisse**), e.g. **marrons** and **yabbies**. (*See also* **langoustine**.)

cream (1) Thick fatty part of milk. Light or reduced cream has a lower fat content. Thickened cream has a thickening agent (e.g. **gelatine**) added to enhance whipping.
(2) Food resembling pure cream in consistency; dessert containing cream; creamed soup, e.g. butter cream, custard cream, cream of mushroom soup. (*See also* **crème** etc., *and below*.)

cream cheese Soft, spreadable cheese.

cream, clotted Cream, skimmed from **scalded** milk, cooled and traditionally served with **scones** and jam for cream tea. (*See also* **Devonshire tea**.)

cream, soured Thick **cream** to which a culture has been added to give a sharp taste. (It is not cream that has gone 'off'.)

crema (*Ital.*) [kray-ma] The thick creamy head on an espresso coffee. (Literally 'cream'.) (*See* Chapter 16.)

crème (*Fr.*) (fem.) [crem] **Cream**, custard.

crème, à la (*Fr.*) (fem.) [ah lah crem] Cooked with **cream**, cream sauce or **custard**.

crème anglaise (*Fr.*) [crem on-glaze] **Custard** of a pouring consistency cooked slowly over a **bain-marie**. (Literally 'English cream'.)

crème brûlée (*Fr.*) [crem broo-lay] Thick, smooth **custard** covered with a thick crust of sugar that has been grilled until **caramelised** and crisp. (*Brûlée* means 'burnt'.)

crème caramel (*Fr.*) [crem cara-mel] **Custard** set in a mould containing a **caramel** sauce. Once turned out, the sauce sits on top and slips down the sides.

crème chantilly (*Fr.*) [crem shahn-tee-yee] Fresh **cream** whipped and sweetened with **vanilla sugar**. (Chantilly is a town near Paris.)

crème chiboust (*Fr.*) [crem shiboost] **Custard** similar to **crème pâtissière** but with **gelatine** added. The eggs are separated and the whipped whites are added to the warm custard. (Chiboust was a pastry-cook in 19th-century Paris.)

crème d', de ... (*Fr.*) [crem duh] Cream **liqueurs** (*crème d'*amandes, *crème de* cacao, *crème de* menthe).

crème de cacao [crem duh kakau] **Liqueur** made from **vanilla** and cocoa beans. There are white and brown styles. (*Cacao* is French for 'cocoa'.)

crème de menthe (*Fr.*) [crem duh mah(n)th] Peppermint-flavoured **generic liqueur**, usually bright green. (*Menthe* means 'mint'.)

crème fouettée (*Fr.*) [crem foo-ettay] Whipped **cream**.

crème fraîche (*Fr.*) [crem fraysh] Thick **cream** to which a culture has been introduced resulting in a sharp taste and almost solid consistency. (Literally 'fresh cream'.) (*See also* **mascarpone**.)

crème pâtissière (*Fr.*) [crem paht-ees-see-yair] Thick **custard** thickened with flour to which beaten egg white is added to provide a **mousse**-like texture that will hold its shape in cakes and desserts. (*Pâtissier* (fem. ~ *ière*) means 'pastrycook' or 'confectioner'.) (*See also* **crème chiboust**.)

creole [cree-yoal] In the style of the West Indies and S USA. Typically, savoury dishes contain **capsicums** or bananas, and sweet dishes contain sugar, **rum** or bananas. (French, *créole* [cray-oal].) (*See* **Cajun**.)

crêpe (*Fr.*) (fem.) [krep] Very thin **pancake** cooked on both sides, usually filled and rolled up when served. *Crêpes* can be sweet or savoury.

crêpes Suzette (*Fr.*) (fem. pl.) [krep soo-zet] Sweet **crêpes** flavoured with orange juice and orange **liqueur**, e.g. **curaçao** or **Grand Marnier**. They are often cooked at the table in a **chafing dish** and served **flambé** in lighted **rum**. (Invented by chef Henri Charpentier *c.* 1900, and named in honour of a child dinner guest of the Prince of Wales.)

crépinette (*Fr.*) (fem.) [krep-in-ett] Finely minced meat with seasonings cooked in a caul (thin membrane); a flat sausage.

cress (1) Salad greens from the mustard family. (2) *Lepidium sativum,* a tiny green seedling with little black seeds commonly known as garden cress or mustard-and-cress. (*See also* **rocket** *and* **watercress**.) (French, *cresson*.)

croissant (*Fr.*) (masc.) [kwa-sahn] Crescent-shaped sweet roll, usually made of **puff pastry**, commonly served at breakfast with butter and jam or chocolate. (Literally 'crescent'.) (*See also* **pain au chocolat**.)

croquembouche (*Fr.*) [crockem-boosh] Conical tower of small **choux** buns filled with **custard** or **cream** and coated with spun sugar, traditionally served at weddings. (Literally 'crisp in the mouth'.)

croque-monsieur (*Fr.*) (masc.) [crock muss-yure] Toasted sandwich, esp. one filled with ham and **gruyère** cheese. (*Croque-madame* has a fried egg on top.) (*Cro quant(e)* means 'crunchy' from *croquer* 'to crunch', 'to devour'.)

croquette (*Fr.*) (fem.) [kro-ket] Minced mixture, usually meat or fish, with vegetables and seasoning, prepared with a sauce to bind and shaped into a small ball, tube or **patty**. It is rolled in egg and breadcrumbs before being fried.

cross-contamination Spoiling of food by germs passing from one organism to another.

crostino (*Ital.*) (*pl.* **crostini**) [cross-teeno/ee] **Canapé** or small savoury toast served as an appetiser. (*Compare* **croûton**.)

croustade (*Fr.*) (fem.) [kroo-stahd] Case of pastry, or deep-fried bread, mashed potato or rice containing a savoury mixture in a cream sauce or a **purée**, e.g. *croustades à la* **marinière**. (*See also* **vol-au-vent**.)

croûte (*Fr.*) (fem.) [kroot] (1) Pastry case. (2) Slice of fried bread on which **appetisers** are served. (*See also below*.)

croûte, en Cooked in a pastry case, e.g. *pâté en cro Cite*, **poulet** *en cro Cite*. (*See* **croûte** *above*.)

croûton (*Fr.*) (masc.) [kroo-ton] Small cube of fried or toasted bread used to **garnish** soups and some salad and vegetable dishes. (*See also* **crostino**.)

crudités (*Fr.*) (fem. pl.) [kroo-di-tay] Bite-sized pieces of vegetable served as an **appetiser** with an accompanying savoury dip. Typical vegetables are carrot, celery, mushroom, cauliflower, broccoli and **capsicum**. Many sauces are used: avocado dip, **mayonnaise**, **taramasalata** or any with a soft consistency.

cruets Salt and **pepper** set, and (now rarely) containers for oil and **vinegar**.

crumb down To brush the tablecloth between courses thus removing all debris. A table is usually crumbed down after the main course and before **dessert**. (*See* Chapter 11.)

crumble A **pudding** with a crumbly mixture of flour, sugar and fat forming a layer over fruit, e.g. apple crumble.

crumpet A light yeast cake, toasted and served with butter at tea.

crust Wine term. Sediment in a bottle of wine.

crustacean(s) Shellfish, including **crayfish**, **lobsters**, **scallops**, prawns, shrimps and crabs. (French *les crustaces*.)

cuisine (*Fr.*) (fem.) [kwi-zeen] (1) Kitchen or kitchen staff, e.g. *chef de cuisine*. (2) (Style of) cookery or food, e.g. French cuisine, Thai cuisine, **cuisine naturelle**.

cuisine naturelle (*Fr.*) [kwi-zeen natoor-ell] Style of cooking using fresh ingredients prepared so that the natural flavour is evident. Fats, alcohol, salt and sugar are used sparingly. (Created by Anton Mosimann, *b.* 1947, at the Dorchester Hotel in London.) (*See also* **haute cuisine** *and* **Pritikin**.)

Cumberland sauce Sweet-and-sour sauce made of orange and lemon juice and zest, redcurrant jelly and **port**, usually served cold with meat. (From Cumberland, county in N England.)

cumin Piquant slightly bitter-tasting seeds of an aromatic plant (*Cuminum cyminum*) widely used in cookery esp. to flavour Indian vegetable curries, Mexican chilli dishes and some eastern European breads.

curaçao [cure-a-soh] Orange **liqueur**, either white or blue in colour. (Originally made with oranges from Curaçao, W Indian island.) (*See also* **triple sec**.)

curd Rich creamy fattening part of milk that can be separated from the watery part or whey.

cure To **preserve** food by drying, salting, **pickling** or **smoking**. (*See also* **brine**.)

curry (*Anglo-Ind.*) Dish of meat or vegetables prepared with spices, particularly **turmeric**; general term for Indian dishes. The English word 'curry' comes from a Tamil (S Indian) word meaning sauce, but the word is little used in Indian cookery.

custard (1) Thick, rich pouring sauce made of milk cooked with eggs and sugar. (*See also* **crème** *and* **crème anglaise**.)

(2) Cold cream or pudding made with milk, eggs and sugar that sets when refrigerated, e.g. **crème brûlée**, **crème caramel**, **crème chiboust** *and* **crème pâtissière**.
(3) Mixture of milk and eggs to which flavourings are added and used as a base for savoury pies, e.g. **quiche Lorraine**.

custard apple Soft fruit of W Indian tree (*Anona reticulata*) shaped like large apple with flesh tasting like **custard**.

cutlet (1) Thick slice of meat on the bone from the neck of lamb or **veal**. (*See also* **papillote**.) (2) Thick slice of fish on the bone. (*See also* **darne** *and* **tronçon**.)

cuvée (*Fr.*) (fem.) [koo-vay] Batch of wine or a blend of wines from different vats.

dahl/dal (*Ind.*) [dahl] Thick **purée** of **lentils**. With **roti** it is the staple food of N India.

daikon radish [dye-ko(n)] Giant white radish which is **pickled** and served beautifully sliced or shredded as a **garnish** with many Japanese meals. (*See also* **horseradish** *and* **wasabi**.)

damper (*Austral.*) Dough of flour and water, traditionally baked in the ashes of an outdoor fire.

Danish pastry Flaky yeast cake topped with icing, nuts and fruit and usually baked in a scroll shape. (Also known as Apple Danish, Apricot Danish, etc.)

dariole (*Fr.*) (fem.) [da-ree-ole] (1) Small steep-sided mould used in making small pastries and **puddings**.
(2) Food that is baked or set in a *dariole*.

Darjeeling tea Fine black Indian tea. (Darjeeling is a hill town in the tea-growing area of NW India.)

darne (*Fr.*) (fem.) [darnuh] Steak or **cutlet** of large round fish such as schnapper or cod, cut through the bone. (*See also* **tronçon.**)

dashi (*Jap.*) **Stock** made with konbu (dried seaweed), katsuobushi (shaved bonito fish) and water.

daube, en (*Fr.*) [ahn dobe] Cooked slowly in a pan; braised, e.g. **boeuf en daube.** (*Une daubière* is 'a stewpan'.)

de (*Fr.*) (masc.) [duh] Of, as in **côte** *de* **boeuf** ('side *of* beef'), **pâté** *de* **foie** ('pâté *of* liver'); spelt **d'** when next word starts with a vowel, as in **mignonnettes** *d'***agneau** ('mignonnettes *of* lamb').

débarrasseur/se (*Fr.*) [day-barrass-ur/urze] Junior or trainee waiter who clears dirty dishes, etc. (From *débarrasser,* 'to clear'.)

decaffeinated coffee/tea Coffee or tea from which **caffeine** has been extracted. (*See* Chapter 14.)

decant To pour a **liquor,** esp. wine, carefully, without raising sediment, from a bottle into a **decanter, carafe** or jug.

decanter Narrow-necked glass jug, usually with a stopper, used for storing or serving wines or **spirits.**

deglaze To remove fat from a pan in which meat has been cooked by pouring alcohol into the pan and boiling briskly. **Stock** or **cream** is then added to form a **jus** or sauce. (French *déglacer.*)

dégustation (1) Tasting; sampling food and drink.

(2) Wine-tasting. (French *dégustation* (fem.) [day-goo-stah-seeo(n)]; Italian *degustazione* [day-guss-tats-ee-ohnay]). (*See also* **menu dégustation** *and* Chapter 2.)

déjeuner (*Fr.*) [day-zjer-nay] Lunch.

de la (*Fr.*) (fem) [duh la] Of the, when used with a feminine noun, as in *de la* **maison** ('*of the* house'). Note that **du** is the masculine form of *de la,* used with masculine nouns.

deli/delicatessen Shop selling cheeses, cooked meats and groceries; casual restaurant. (Deli is short for delicatessen, literally 'delicacies' in German.) (*See also* **charcuterie.**)

délice (*Fr.*) (masc.) [day-leece] Trimmed and folded fillet of fish.

demi-glace (*Fr.*) (fem.) [demee gluss] Rich brown sauce made by **reducing stock** to half its original volume. **Madeira, sherry** or **brandy** is often added after the reduction. (Literally 'half glaze'.)

demitasse [demee tass] Very small **coffee** cup for after-dinner and short black coffee. (*See also* **Greek, Turkish** *and* Chapters 3 *and* 16.) (Literally 'half cup' in French.)

des (*Fr.*) (*pl.*) [day] Of, when used as a plural, e.g. **purée** *des* **marrons.**

desiccate To preserve by drying, e.g. desiccated **coconut.**

dessert (1) Final, **sweet** course of a meal. (2) Fruit and nuts served at the end of a meal. (*See* Chapter 2 *and* **classic menu.**)

dessert wine Sweet wine usually served with dessert. (*See* **auslese, muscat, spätlese, sauternes** *and* Chapter 14.)

devilled Marinated meat in a thick, slightly sweet, spicy sauce. A devil mixture usually contains one or all of the following: **mustard, Worcestershire sauce,** sugar and tomato sauce. (*See* **diable**.)

Devonshire tea Afternoon **tea** with **scones, clotted cream** and jam. (From Devon, county in SW England.)

diable, à la (*Fr.*) [ah lah dee-ah-bluh] **Devilled,** as in **boeuf** *à la diable*. (*Diable* means 'devil'.)

dice To cut up into small even cubes. (*See also* **macédoine**.)

digestif (*Fr.*) (*masc.*) [dee-zjes-teef] A digestive; an after-dinner drink, particularly a **liqueur**, said to aid digestion, whence the name.

Dijon (*Fr.*) [dee-zjo(n)] **Mustard** that is prepared with **verjuice** and white wine. (Dijon is a city in Central W France.)

dill Fragrant annual plant, the feathery leaves of which are used as a **garnish** and as a flavouring herb, esp. with fish. The seeds are used in **pickles**. (*See* **dill pickle** *below; see also* **gravlax**.)

dill pickle Baby cucumber pickled with **dill** seeds.

dim sim/sum (1) (*Chin.*) Small savoury items, e.g. **won tons**, served at **yum cha**.
(2) (*Austral.*) Small savoury **dumpling** containing meat and cabbage wrapped in a **won ton** wrapper and steamed or deep-fried.

distillation Process used to make **spirits** from **fermented** 'wine'. (*See* Chapter 14.)

doily/doyley Lace-like paper mat used to decorate plates, particularly under cakes, or on an **underliner**.

dolcelatta (*Ital.*) [doll-che-lah-tah] Moist, creamy, mild, blue-veined cheese made from cow's milk. (Literally 'sweet milk'.)

dolci (*Ital.*) (*pl.*) [doll-chy] Sweets; desserts. (*See also* **primi piatti** *and* **secondi piatti**.)

dolmas (*Grk & Turk.*) [doll-mahth] (*pl.* **dolmades** [doll-mah-thess]) Vine (or cabbage) leaves stuffed with oily rice, sometimes with nuts, currants and minced lamb, etc. in the rice. Dolmades are often served with **meze**.

döner kebab (*Turk.*) [doaner kuhbahb] **Marinated** lamb, thinly sliced from a vertical **salamander**, rolled with salad, **tahini** and **hummus** in **pita** bread. (Lebanese equivalent is **chawarma**.)

doppio (*Ital.*) Double shot of espresso coffee; extra strong drink of coffee. (*Doppio* means 'double'.)

doyley *See* **doily**.

Drambuie Proprietary **liqueur** made from Scotch **whisky**, heather, honey and herbs.

draught beer Beer drawn from a barrel (keg) rather than bottled or canned.

dredge To sprinkle with castor sugar, flour, etc.

dressing *See* **French dressing** *and* **vinaigrette**.

dry ginger (ale) Non-alcoholic drink of carbonated water flavoured with ginger extract. Often called 'dry' as in 'brandy and dry'. (*See also* **ginger**.)

Dry Martini Cocktail of gin with a little dry (French) **vermouth**, **garnished** with a green **olive** or a twist of lemon.

du (*Fr.*) (masc.) [doo] Of the, when used with a masculine noun, as in **carte** *du jour* ('menu *of the* day'). Note that **de la** is the feminine form of **du**; **des** is the plural.

Dubonnet French brand of aromatised wine or **vermouth** commonly served with **soda** or with ice as an **apéritif**.

duchesse *See* **pommes duchesse**.

du jour (*Fr.*) (masc.) [doo zjoor] Literally 'of the day', e.g. **poisson** *du jour* ('fish of the day').

dukkah (*Egypt*) Crushed nuts, esp. hazelnuts, blended with spices and eaten with bread dipped in olive oil.

dumb waiter (1) A small lift for moving food, etc. between the kitchen and the dining-room.
(2) A table-top stand with a revolving top, a **lazy Susan**.

dumpling Ball of dough, either baked or simmered in a **casserole**; dumplings can be sweet (flavoured with sugar, syrup, etc.) or savoury (containing herbs and spices). (*See also* **dim sim**, **gnocchi**, **quenelle** *and* **won ton**.)

duxelles (*Fr.*) (fem. pl.) [dook-sel] **Sautéed** mixture of finely chopped onions or **shallots** with mushrooms, seasoned and **reduced**. Duxelles are used as a **garnish** or stuffing. Such dishes are described as à la *duxelles* (as in **omelette** *à la duxelles*).

Earl Grey tea Smoky-tasting blended **tea** flavoured with **bergamot**. (Earl Grey, 1764–1845, was a British statesman and reforming Prime Minister, 1830–4.)

échalote (*Fr.*) (fem.) [ay-shallot] **Shallot**.

éclair (*Fr.*) (masc.) [ay-clair] Small cylindrical cake made from **choux** pastry, split and filled with **cream** and iced with chocolate.

écrivisse (*Fr.*) (fem.) [ay-cree-veece] European freshwater crayfish, used also as the French word for **yabby**. (*See* **crayfish**.)

EFTPOS Electronic funds transfer at point of sale (system) (*See* Chapter 17.)

egg-nog/egg flip Alcoholic drink of liquor mixed with eggs and milk served either just warm or cold.

eggs Benedict Poached eggs with grilled ham served on toasted bread or **muffin** and covered in **hollandaise**.

émincé (*Fr.*) (masc.) [ay-mah(n)-say] Thin slice, esp. of meat; rasher (of bacon). (*Emincer* means 'to slice finely'.) (*See also* **émincés** *below*.)

émincés (*Fr.*) (masc. pl.) [ay-mah(n)-say] Thin slices of left-over meat reheated in **jus** or sauce. (*See also* **émincé** *above, and* **réchauffé**.)

en (*Fr.*) [ah(n)] In, as in *en* **cocotte**. (*See also* **chevreuil**, **cocotte**, **croûte**, **daube**, **gelée** *and* **papillote**.)

enchilada (*Mex.*) [en-chill-ahda] Fried **tortilla** filled with cheese and/or meat and rolled. Served with **chilli** sauce and, possibly, **guacamole**.

endive [en-dive] (1) *Cichorium endivia*, sometimes referred to as curly endive, a bitter green-leaf vegetable used in salads. Also known as **frisée**.
(2) (esp. *USA*) Belgian endive, **chicory** (*Cichorium intybus*) or **witloof**, used in salads or as a cooked vegetable. (Both meanings may be encountered in Australia but the first is more common.)

English Breakfast tea Blended Indian and Ceylon (black) **teas**.

en pension (*Fr.*) [ah(n) pah(n)-see-ah(n)] Full board, with all meals provided.

ensalada (*Span.*) Salad.

entrecôte (*Fr.*) (masc.) [ahn-truh-coat] Rib steak of beef, often pounded and served as **minute steak**. (Literally 'between the rib'.) (*See also* **filet mignon, porterhouse, rump, sirloin** *and* **T-bone.**)

entrée (*Fr.*) (fem.) [ahn-tray] In Australia, the first **course** of a meal. In the USA (and often in the UK), it refers to the main dish of the meal. In the **classic menu**, the entrée was the first (and smallest) of the meat courses, following the fish course. (Literally 'entrance'.) (*See* Chapter 2.)

entremets (*Fr.*) (masc. pl.) [ahn-truh-may] Sweet **course**. (*See* **classic menu.**)

épaule (*Fr.*) (fem.) [ay-pole] Shoulder, e.g. *épaule d'*agneau, 'shoulder of lamb'.

épinards (*Fr.*) (masc. pl.) [ay-peenard] Spinach. (*See also* **silverbeet.**)

escabèche (*Fr.*) [eska-baish] Fried fish preserved in a **sweet-and-sour dressing** containing vinegar; fried fish marinated in **vinaigrette**. (From Spanish *escabeche* 'pickle'.)

escalope (*Fr.*) (fem.) [ess-cal-op] Thin slice of boneless meat, e.g. *escalope de* veau.

escargot (*Fr.*) (masc.) [ess-car-go] Snail.

eschalot Shallot.

espagnole (*Fr.*) [ess-pa(n)yol] Spanish, as in **omelette farcie** *à la* espagnole ('Spanish omelette'). (*See also below.*)

espagnole, sauce (*Fr.*) (fem.) [sohs ess-pa(n)yol] **Reduced** brown sauce flavoured with **tomatoes** and **sherry**.

(*See also* **salmis, sauce.**)

espresso (*Ital.*) [ess-press-oh] *See* Chapters 14, 15 *and* 16.

essence Concentrated extract from fruit, flower or nut used for flavouring, e.g. **vanilla essence**.

establishment A business in the hospitality industry—restaurant, hotel, etc.; a hospitality operation.

extra virgin *See* **virgin olive oil.**

fajita (*Mex.*) [fa-heetah] Mexican corn bread; grilled meat or chicken served in a soft **tortilla** with tomato and corn **salsa**, **chilli** sauce and cream.

falafel/felafel (*Arab.*) [fell-ah-ful] (1) **Chick-pea** ball flavoured with spices. (2) Roll of **pita** bread filled with salad, **hummus** and **chick-pea** balls.

farce (*Fr.*) (fem.) [farse] Savoury stuffing; **forcemeat**.

farci(e) (*Fr.*) [farsee] Stuffed; a stuffed dish. (*Farcir* means 'to stuff'.)

farinaceous [farin-ay-shus] Consisting of cereals, beans or pulses. (*Farina* is Latin for flour.)

farro (*Ital.*) Flour made from traditional grain used to make bread and **pasta**. (Also known as **spelt**.)

fattoush (*Arab.*) Spicy salad containing toasted **pita** bread. (*Compare* **panzanella.**)

felafel *See* **falafel.**

fennel (1) Large bulbous plant that tastes of aniseed. It is **braised** and served as a vegetable or can be sliced raw and mixed with salad greens.
(2) Feathery fronds of the fennel plant used as a herb or **garnish**, esp. with

fish.

(3) Aromatic seeds used as a flavouring, esp. in **pickled** foods.

fermentation Process of converting a liquid containing sugar into alcohol by the action of yeast.

Fernet Branca [fair-nay brankah] Brand of Italian **bitters**, usually served as an **apéritif** with **soda**.

feta/fetta (*Grk*) Cheese, traditionally made from ewe's milk, preserved in **brine**. It is a solid white cheese with a sharp, salty taste. (*Fetes* means 'blocks' or 'slices', the curd being cut into blocks before being preserved.)

fettuccine (*Ital.*) (*pl.*) [fett-u-cheeny] Thin ribbons of **pasta** cut to a length of about 30cm. Very similar to **tagliatelle**. (*Fettuccia* means 'ribbon'.)

feuilleté (*Fr.*) [foo-ee-et-ay] **Puff pastry** case, traditionally cut into a leaf shape or triangle, e.g. *feuilleté de* **coquilles**, scallops cooked in a puff pastry case. (Literally 'foliated' or 'leaflike'. *Une feuille* is 'a leaf', but nowadays many shapes are referred to as '*feuilletés*'.) (*See also* **pâte feuilletée**.)

ficelle, boeuf à la *See* **boeuf à la ficelle**.

filet (*Fr.*) (*masc.*) [fill-ay] **Fillet**, e.g. *filet de* **boeuf**, 'fillet of beef'.

filet mignon (*Fr.*) (*masc.*) [fill-ay mee-neeyo(n)] Small, round **fillet** steak. (*See also* **entrecôte, minute steak, porterhouse, rump, sirloin** *and* **T-bone**.)

fillet (1) (noun) Thick boneless piece of meat or fish, usually the prime cut, e.g. fillet steak.
(2) (verb) To remove from the bone, e.g. filleted fish.

filo pastry [feel-oh] Very thin paper-like pastry, buttered or oiled, and cooked in layers.

fine champagne Description of the finest kind of **cognac** (i.e. **brandy**, *not* **sparkling wine**).

fine dining (room) High-class dining (room) with formal service, typically found in five-star hotels.

fines herbes (*Fr.*) (fem. pl.) [feen airb] Herbs (**chervil**, parsley, **tarragon** and **chives**) chopped up and mixed together, used to flavour food, typically **omelettes**, meat and various sauces.

fingerbowl Small bowl filled with water (and perhaps a piece of lemon) placed on the table so that guests can clean their fingers.

finish Wine-tasting term. The last taste impression from a mouthful of wine.

fino [fee-noh] Style of Spanish **sherry**.

flambé(e) (*Fr.*) [flahm-bay] Flamed. Food served *flambé* has had **spirits** poured over it and then been ignited. (*See* Chapter 12.)

flan Open tart containing a filling (of **custard**, fruit, etc.) (*See also* **frangipane**.)

flatware (1) (*UK*) plates, saucers, crockery.
(2) (*Austral.* & *USA*) cutlery.
(*Compare* **hollowware**.)

flat white *See* Chapter 16.

fleuron (*Fr.*) (*masc.*) [flur-o(n)] Crescent- or flower-shaped edible decoration (e.g. a *fleuron* of **puff pastry**).

float (1) (verb) Cocktail-mixing term; to place one liquid above another in the glass so that they do not mix but one 'floats' above the other.

(2) (noun) Change placed in the till before trading begins.

floating islands Soft balls of **meringue** poached in milk and served in **custard** sauce. (In French, **oeufs à la neige.**)

florentine [flo-ren-teen] Thin biscuit of **glacé** fruit and nuts coated on one side with chocolate, often served after dinner with coffee.

florentine, à la (*Fr.*) [ah lah flo-ren-teen] In the style of Florence (Italian city). Cooked or served with spinach and (usually) **mornay** sauce, e.g. **suprêmes** of chicken *à la florentine.*

flute (1) (noun) Narrow-stemmed glass used for serving **sparkling wines.** (*See* Chapters 13 *and* 15.)

(2) (noun) Long thin French roll.

(3) (noun) Groove made on food for decoration.

(4) (verb) To make grooves on the surface of a food item, such as a pie crust or a **mousse.**

focaccia (*Ital.*) [fok-ah-cheea] Flat bread similar to a **pizza** base, often served with **antipasto** or split and made into a sandwich.

foie (*Fr.*) (masc.) [fwah] Liver, e.g. *foie de veau*, 'veal liver'.

fondant Soft mixture of flavoured sugar used as icing on cakes or served as a **sweetmeat,** sometimes coated with chocolate.

fondue (*Swiss Fr.*) (fem.) [fon-doo] Dish cooked at table by dipping small pieces (usually of meat or bread) into a bowl containing hot oil or melted cheese. (*Fondu* is French for 'melted'.) (*See also* **raclette.**)

food allergy An immunological reaction to food proteins. (*See also* **food intolerance** *below.*)

food intolerance A pharmacological reaction (like side effects from a drug) to the chemicals in food. (*See also* **food allergy** *above.*)

fool Cold dessert made of **puréed** fruit and whipped **cream.**

forcemeat Meat chopped and seasoned for use as a stuffing; a **farce.**

forestière, à la (*Fr.*) [ah lah foress-tee-air] Served with **sautéed** mushrooms. (*Forestière* means 'forester's wife' or 'female forester'.)

formaggio (*Ital.*) Cheese.

fortified wine Wine strengthened by the addition of some **spirit.** Fortified wines include **muscat, sherry** and **port.** (*See* Chapter 14.)

four, au (*Fr.*) [oh foor] Oven-baked; roasted. (*Four* means 'oven'.)

framboise (*Fr.*) (fem.) [frahm bwaz] Raspberry.

française, à la (*Fr.*) [ah lah frahn-sez] In the French style; usually served with **demi-glace** sauce. **Petits pois** *à la française* are cooked with lettuce and onion.

frangipane [fronzji-pahn] **Custard** made with ground almonds or crushed **macaroons.** (In French, *crème frangipane.*) (*See also* **crème pâtissière.**) (Named after an Italian marquis, Muzio Frangipani, who lived in Paris in the 16th century and devised a bitter-almond scent for sprinkling on gloves.)

frangipane tart/flan Jam-lined pastry case filled with **frangipane** cream and decorated with fresh **cream** and **pistachio** nuts.

frankfurter Smoked beef-and-pork-sausage, usually simmered and served in a bread roll as a hot dog. Tiny **cocktail** frankfurters are served as **appetisers**. (Frankfurt is a city in central Germany.)

frappé (*Fr.*) [frappay] (1) (Food) served on crushed ice.
(2) Cocktail-mixing term; poured over crushed ice, e.g. **crème de menthe** *frappé*. (Literally 'crushed' or, more exactly, 'beaten'.)

free-range Allowed to roam and forage for food. The term most commonly refers to chickens or hens and their eggs, but is also used to describe the meat from animals reared out of doors, esp. pork.

French Dry white **vermouth** (as in **gin-and-French**).

French dressing Cold dressing for salad, etc. made with oil and **wine vinegar**, and seasonings such as **pepper**, salt, **garlic** and **Dijon mustard**. Lemon or verjuice is sometimes used instead of vinegar. (*See* **vinaigrette**.)

French toast Bread dipped in beaten egg and fried.

freshly squeezed (juice) Juice, esp. orange, squeezed on the premises from fresh ingredients (i.e. not served from a bottle of juice, even if it is labelled 'freshly squeezed').

friande [free-ah(n)d] Small oval-shaped cake made with ground almonds.

friandises (*Fr.*) (fem. pl.) [free-ah(n)-deez] Dainties, e.g. fruits dipped in chocolate, or **petits fours**, often served at the end of a meal with the **coffee**.

fricassee [frik-a-see] Pieces of meat, usually a white meat such ase chicken, cooked in **stock** and served in a white sauce. Fricassee has been adopted as an English word; the French is *fricassée* [frik-a-say]; so you should say 'chicken fricassee' (pronounced *see*) but *fricassée* (pronounced *say*) *de* **poulet**.

frijole (*Span.*) [free-hoh-lay] Kidney bean. (*See also* **refried beans**.)

frill Fancy paper decoration placed on cutlet or ham bones, etc. (*See also* **papillote**.)

frisée (*Fr.*) [free-zay] **Curly endive** (*Cichorium endivia*); bitter green-leaf salad vegetable. *Frisé(e)* is French for 'curly'.

frite (*Fr.*) [freet] Fried. (*See* **pommes frites**.)

frittata (*Ital.*) [frit-ahta] **Omelette**; beaten eggs fried on both sides to form a cake. Vegetables are sometimes beaten into the egg mix. (*Fritto* means 'fried'; *see* **fritto misto**.)

fritter Food, either sweet or savoury, dipped in **batter** and deep-fried. (*See also* **pakora**.)

fritto misto (*Ital.*) Mixed vegetables deep-fried in **batter**. (Literally 'fried mixed'.)

fritto misto de pesce (*Ital.*) [fritoh mistoh dee pescay] Mixed **seafood** deep-fried in **batter**. (Literally 'fried mix of the sea'.)

froid (*Fr.*) [fwah] Cold, as in **buffet froid**.

fromage (*Fr.*) (masc.) [from-ahje] Cheese.

fromage blanc (*Fr.*) [from-ahje bla(n)] Low-fat fresh cheese made from skimmed milk. It is used in place of cream, or as an accompaniment to fruit.

frontignac [fron-ti-nyak] Grape variety. (*See* **muscat**.)

front of house The part of the **establishment** seen by the guests; the dining-room as opposed to the kitchen. (*See also* **back of house**.)

frost (1) To cover with sugar, esp. a cake. Frosted fruit is dipped in egg white or lemon juice so that the sugar will cling to it. (*See also* **frosting**.)
(2) To garnish a glass by dipping the rim in lemon juice and sugar or salt.

frosting (*USA*) Soft icing used to decorate cakes.

fruits (*Fr.*) [froo-ee] Fruit.

fruits de mer (*Fr.*) (masc. pl.) [frooee duh mair] Seafood. (Literally 'fruits of the sea'.)

fruity Wine-tasting term. Pleasing taste of a young wine; description of its **aroma**.

Fukien (*Chin.*) [foo-kyen] One of the five main styles of Chinese cuisine. Fukien cuisine is known for its **clear soups** and **seafood**. (Fukien, now called Fujian, is a coastal province of China.) (*See also* **Cantonese, Honan, Peking** *and* **Szechwan**.)

fumé (*Fr.*) [foomay] (1) **Smoked**.
(2) Style of white wine, usually a **sauvignon blanc**, cloudy and piquantly tasty, called fumé blanc.

function Prearranged occasion (banquet, dinner, cocktail party, etc.) for a known number of guests.

funghi (*Ital.*) (*pl.*) (sing. **fungo**) [foongee] Mushrooms, e.g. **risotto** *alla funghi*.

gado-gado (*Indon.*) [gahdo gahdo] Vegetables cooked in peanut sauce with hard-boiled egg.

galantine [galan-teen] Boned meat, usually stuffed with **forcemeat**, pressed into shape and covered in **aspic** jelly. It is served cold. (*See also* **brawn**.)

galette (*Fr.*) (fem.) [ga-let] (1) Flat cake; **pancake**.
(2) Thick shortbread biscuit.

Galliano Golden-yellow herb **liqueur** from Milan (city in N Italy). (Named after Major Galliano, a hero of the Italian-Abyssinian war of 1896.)

game All wild animals and birds that are hunted. (In French *gibier*.) (*See also* **venison**.)

gammon Ham, usually served hot as gammon steak.

ganache (*Fr.*) [gan-nash] Chocolate sauce made with pure chocolate (**couverture**) and cream, used as a coating or filling for cakes, biscuits, etc. (*See also* **Valrhona**.)

garam masala (*Ind.*) [garram ma-sah-la] Ground spices used as a flavouring in curries, etc. Different mixtures are used for different purposes. (*Garam* means 'hot'.) (*See also* **curry** *and* **masala**.)

garbure (*Fr.*) (fem.) [gar-boor] A very thick, hearty soup containing cabbage, beans and **pickled** meat (typically **confit d'oie**), served ladled over wholemeal bread. (*See also* **potage** *and* **soupe**.)

garde-manger (*Fr.*) (masc.) [gard marnge-ay] Specialist cook in charge of cold items and decorative work in a large kitchen.

garlic Strong-smelling flavouring herb from the same family as onions. The garlic bulb consists of segments called cloves.

garni(e) (*Fr.*) (adj.) [gar-nee] Decorated, e.g. **bouquet garni**, *choucroute garnie* (garnished **sauerkraut**).

garnish (1) (noun) Items placed on or around a dish (or drink) for decoration or taste.
(2) (verb) To place such items on a dish or in a drink.

garniture Items in a **garnish**; the trimmings.

gastronome An expert on good eating and drinking; an epicure.

gastronomy Science or art of good eating and drinking.

gâteau(x) (*Fr.*) (masc.) [gah-toe] Cake(s).

gazpacho (*Span.*) [gaz-pah-cho] Cold soup of **puréed tomatoes**, cucumber, onions, **garlic** and **red pepper**, served with condiments, e.g. olives, hard-boiled egg and **croûtons**.

gear Cutlery, e.g. '**dessert** gear' (the spoon and fork for the dessert), '**service gear**' (a serving spoon and fork). (*See* Chapters 4 *and* 8; *see also* **flatware** *and* **hollowware**.)

gelatine/gelatin [jell-a-tin] Almost colourless, odourless, tasteless, soluble substance extracted from animal skin, bones and cartilage, etc. Used to make jellies and various cold desserts. (*See also* **aspic**.) (NB Many vegetarians prefer not to eat gelatine.)

gelato (*Ital.*) [jel-ahtoh] Ice cream made with less milk than **semifreddo** and no egg yolk. As a result, it is much lighter in texture. (*See also* **sorbet** *and* **water ice**.)

gelée (*Fr.*) (fem.) [zjelay] Jelly.

gelée de cuisine [zjelay duh kwi-zeen] **Aspic** jelly.

gelée, en [ah(n) zjelay] In jelly, e.g. **oeufs poché** *en gelée*.

generic Of a general type (which anyone may make), as opposed to **proprietary** (particular brand, which is the exclusive property of a person or business); also as opposed to **varietal** in reference to wine. (*See* Chapter 14.)

genever [jenny-vur] Style of Dutch gin. (It has no connection with Geneva, the Swiss city, but is derived from the French word for 'juniper', *genièvre*.)

génoise (*Fr.*) (fem.) [gay-nwaz] Genoese sponge; light sponge cake.

gewürtztraminer [guh-voorts-tramin-er] Variety of the **traminer** grape used to produce spicy white wine. (*Gewürtz* is German for 'spicy'.)

ghee (*Ind.*) [gee, with a hard 'g'] Clarified cow- or buffalo-milk butter that doesn't burn. There is a much cheaper vegetable-oil ghee substitute, called *vanaspati* (vegetable) ghee.

gibier (*Fr.*) (masc.) [zjee-bee-ay] **Game** (wild animals and wildfowl). (*See also* **venison**.)

giblets [jib-lets] Innards of poultry and **game** birds, the edible portions of which are the liver, heart, etc.

Gibson Cocktail of gin with a little dry (French) **vermouth**, **garnished** with a pearl (cocktail) onion. (*Compare* **Dry Martini**.)

gigot (*Fr.*) (masc.) [zjee-goh] Leg (of mutton or lamb).

gin Spirit made from grain alcohol distilled with selected herbs and fruit—in particular, juniper berries. The most popular style, used in **cocktails**, is London Dry. The other main style, hollands (or **genever**) gin, has a strongly distinctive flavour that makes it less suitable for mixing.

ginger The root of *Zin giber officinale*, used freshly grated in **curries**, dried and powdered as a sweet or savoury spice, or preserved in syrup and **glacéed**.

ginger ale *See* **dry ginger ale**.

ginger beer Slightly alcoholic, aerated drink made of **ginger, sugar syrup** and yeast.

gingerbread Golden-brown soft biscuits or cake made into shapes and decorated with icing, sweets, etc., e.g. gingerbread man, gingerbread house.

Ginger Wine, Green *See* **Green Ginger Wine**.

Gippsland blue Fine **blue-veined** cheese from Gippsland in SE Victoria.

girella Boneless cut of lamb or veal from the top of the leg.

glace (*Fr.*) (fem.) [gluss] (1) Ice; ice cream; iced drink.
(2) **Glaze**, e.g. *glace de* **viande** ('meat glaze').

glacé(e) (*Fr.*) (adj.) [gluss-ay] (1) Iced, icy.
(2) Glazed. Preserved in sugar, e.g. glacé fruits.

glaze To make shiny, usually by coating with beaten egg or with a **reduced stock**. (*See also* **miroir**.)

globe artichoke *See* **artichoke**.

glühwein (*Ger.*) [gloo-vine] Spicy **mulled wine**.

gluten Elastic protein substance found in wheat flour. Sticky when wet, it

holds the dough together and traps the air bubbles produced by the **fermenting** yeast, so helping the bread to rise. (NB Some people with coeliac disease cannot tolerate gluten in their diet.)

gnocchi (*Ital.*) (*pl.*) [nee-o-kee] Small **pasta dumplings** made from potato, pumpkin or similar vegetable, or **semolina**, served in a sauce.

goi cuon (*Viet.*) [goi kuon] Chewy, soft, transparent **rice-paper** rolls containing rice **noodles**, prawns and pork. They are eaten cold dipped in a peanut and bean sauce. (*Goi* means 'raw food', *cuon* is a roll.) (*See also* **cha gio** *and* **spring roll**.)

gomme syrup Sugar syrup.

gorgonzola (*Ital.*) Moist, creamy **blue-veined** cheese made from cow's milk. (Gorgonzola is a village near Milan, N Italy.) (*See also* **Gippsland blue, roquefort** *and* **stilton**.)

gougère (*Fr.*) (fem.) [goo-zjair] Cheese-flavoured **choux** puff served warm. In **Burgundy** they are traditionally eaten cold when wine-tasting.

goujon (*Fr.*) (masc.) [goo-zjo(n)] Thin strip of fillet of fish or chicken (a *goujon* is a gudgeon, a small freshwater fish).

goulash (*Hun.*) [goo-lash] Red stew of meat and onions seasoned with **paprika** and garnished with potatoes.

gourmand (*Fr.*) (masc.) [goor-mond] Lover of food; a glutton. (*Compare* **gourmet**.)

gourmet (*Fr.*) (masc.) [goor-may] Expert in food and wine; an epicure or **gastronome**. (*Compare* **gourmand**.)

grana (*Ital.*) [grah-na] Hard cheese, esp. *grana Parmigiano Reggiano* (**parmesan**), or *grana Padano*. (*See also* **Parmigiano**.)

Grand Marnier [gron marny-ay] Proprietary **liqueur**, **cognac**-based, with an orange flavour.

granita (*Ital.*) [gran-ee-tah] Crushed ice drink flavoured with fruit, coffee or wine, and spices.

grappa (*Ital.*) Italian **spirit** made from the remnants of grapes (skins, stalks, pips, etc.) after the **fermented** juice or **wine** has been removed.

gras (*Fr.*) (adj.) [grah] Fat (as in **pâté de foie gras**).

gras, au (*Fr.*) [oh grah] With ham or bacon.

gratin, au (*Fr.*) [oh grat-i(n)] With a golden crust; sprinkled with breadcrumbs and/or grated cheese, and baked or grilled. (*See also* **gratiné(e)** *below*.)

gratiné(e) (*Fr.*) (adj.) [grat-ee-nay] Cooked, usually with breadcrumbs and grated cheese, to form a golden crust. The word can be anglicised into 'gratinated'.

gravlax (*Scan.*) Fish (properly salmon) marinated with salt and **dill**, served thinly sliced, usually with a mustard sauce.

grazing Practice of eating a succession of modest snacks or small helpings of different dishes, instead of having a substantial meal.

grecque, à la (*Fr.*) [ah lah grek] In the Greek manner; mushrooms and vegetables with **coriander** seeds cooked in olive oil, served cold.

Greek coffee Strong, dark, sweet **coffee** served in small cups; **Turkish** coffee.

Green Ginger Wine Proprietary **fortified wine** made from dried grapes and flavoured with **ginger**.

green pepper *See* **sweet pepper**.

green peppercorn *See* **peppercorn**.

gremolata (*Ital.*) [grem-oh-lahta] Parsley chopped with lemon **zest** and, usually, garlic, traditionally served as a **condiment**, esp. sprinkled over **osso buco**. (*Compare* **persillade**.)

grenadine [grenna-deen] Red pomegranate-flavoured **cordial**, used for sweetening or colouring drinks.

grissino (*Ital.*) (*pl.* **grissini**) [griss-eenoh] Long, thin, crisp bread-stick served with **antipasto**.

gros sel (*Fr.*) (masc.) [grow sell] Rock salt. (Literally 'coarse salt'.)

gruyère (*Swiss Fr.*) [groo-yair] Cow's-milk cheese, firm and smooth in texture with small holes or eyes. The rind isn't usually eaten. (Gruyère is a village in Switzerland.) (*See also* **raclette**.)

guacamole (*Mex.*) [gwah-kah-mo-lay] Spicy **salsa**, chunky in texture, made with avocado, onions, chilli and, sometimes, tomato. (*See also* **enchilada** *and* **nacho**.)

guéridon [geri-don] Trolley or table on which food is prepared in the dining-room. (*See* Chapters 3 *and* 12.)

guéridon service Service from a **guéridon**. (*See* Chapter 10.)

Guinness Famous Irish **stout**, originally brewed in Dublin, but now also elsewhere including (under licence) Australia.

gumbo (*Creole* & *Cajun*) (1) **Okra** or **lady's fingers**.
(2) Spicy **casserole** of **seafood** and vegetables (including **okra**). (*See also* **bouillabaisse** *and* **pochouse**.)
(3) (*USA*) **Okra** soup, or other dishes containing **okra**.

HACCP Hazard Analysis Critical Control Point. HACCP is a system or set of procedures designed to prevent or eliminate food safety hazards.

hâché(e) (*Fr.*) [ah-shay] Minced; mashed, e.g. **bifteck** *hâché* (**minced beef**, hamburger).

haggis (*Scot.*) Traditional dish of oatmeal and chopped **offal** stuffed and sewn into a sheep's stomach and steamed or poached.

halal (*Arab.*) [ha-laal] Lawful for Muslims; meat killed and prepared according to Muslim law. (Muslim equivalent of the Jewish **kosher**.)

haloumi (*Grk*) [haloo-mee] Hard goat's-milk cheese, usually grilled and served with mint and lemon.

halva (*N Africa* & *E Medit.*) Light, crunchy **sweetmeat** of ground **sesame** seeds, sugar, almonds, etc.

haricot (*Fr.*) (masc.) [arry-coh] Bean, esp. the small white navy bean used, for example, in a **cassoulet** or for baked beans. (*See also* **haricot vert**.)

haricot vert (*Fr.*) (masc.) [arry-coh vair] French, green or string (runner) bean. (*Vert* means 'green'.)

harissa (*N Africa*) **Condiment**; fiery red paste made with chillies, spices and, usually, mint.

hash browns (*USA*) Cake of **puréed** potato and onion, bound with egg, and fried.

hasselback potatoes Potatoes sliced almost all the way through, baked with oil or butter and a topping of mixed herbs and **paprika**.

haute cuisine (*Fr.*) (fem.) [ote kwiz-een] French cooking and service of a very high standard. (*Haute* means 'high'.) (*See also* **cuisine**.)

Hawthorn strainer Cocktail strainer; it has two prongs that fit over the side of the blender with a wire coil that fits inside the rim.

herb Aromatic plant used to flavour food, e.g. parsley, **chervil**, **tarragon**. (*See also* **fines herbes**.)

herbal tea An **infusion** made by adding boiling water to the leaves or flowers of various herbs. Most, but not all, herbal teas are **caffeine**-free. Also called **tisane**. (*See also* **ayurvedic tea**, **chai**.)

hermitage (1) (*Fr.*) [air-mit-ahzj] A famous Rhône wine.
(2) (*Austral.*) [herm-it-idge] Formerly an alternative name for the grape variety **shiraz**.

highball (1) Any **spirit** mixed with ice and soda or **dry ginger** served in a highball glass, e.g. a **Whiskey Highball**.
(2) A style of tall, straight-sided glass. (*See* Chapter 13 *and* subsequent diagram of glassware.)

hock English name for German Rhine wine; **riesling**; **generic** wines made in that style. (From Hochheim, German town from which hocks were shipped.)

hoi sin sauce (*Chin.*) Sweet and spicy sauce made of **soya beans**, **garlic** and spices and served as a **condiment**, esp. with **Peking duck** and pork dishes.

hokkien noodles (*Chin.*) Fresh thick **noodles**, either white or yellow in colour.

hollandaise [oll-and-ayz] Rich sauce made from the careful **blending** of egg yolk, butter and lemon juice or **vinegar**. The French term is **sauce** *hollandaise* [sohs oll-and-ayz].

hollands Dutch style of **gin**.

hollowware, hollow-ware Knives and forks; **cutlery**. (*See also* **flatware**.)

homard (*Fr.*) (masc.) [om-ahr] **Lobster**. (*See also* **crayfish, langouste** *and* **rock lobster**.)

hommos *See* **hummus**.

Honan (*Chin.*) [hoe-nan] One of the five main styles of Chinese cuisine. **Sweet-and-sour** dishes are typical of spicy *Honan* cooking. (Honan, now called Henan, is a province in central China.) (*See also* **Cantonese, Fukien, Peking** *and* **Szechwan**.)

hors d'oeuvre(s) (*Fr.*) (masc.) [or-durv] **Appetiser** or small **savoury** served before or at the beginning of a meal. (Literally 'before the work'.)

horseradish Pungent root of the mustard family (*Cochlearia armoracia*). (*See also* **daikon, horseradish sauce** *and* **wasabi**.)

horseradish sauce Grated **horseradish** mixed with **cream, vinegar** and seasonings and traditionally used as a **condiment** with roast beef, **smoked** trout and mackerel.

host(ess) (1) Patrons who are entertaining their own guests in an **establishment** and who are responsible for paying. (*See* Chapter 5.)
(2) Person who greets the restaurant's patrons, takes their coats and shows them to their table—'mine host'.

hot-smoked *See* **smoke**.

house wine Moderately priced wine, usually bought in bulk by an **establishment** and often served from **carafes**. (*See* **vin ordinaire**.)

huître (*Fr.*) (fem.) [ooee-truh] **Oyster**.

hummus/houmos (*Grk* & *Arab.*) [hoomuss] **Purée** of **chick-peas** and **sesame seed** paste, often served with **pita** bread as a **meze** or **hors d'oeuvre**.

ibrik (*Turk.*) Small metal jug (about 30mL) with a long handle and narrow neck used for brewing coffee.

iced coffee *See* Chapters 14 *and* 16. (*Compare* **affogato, caffè frappé** *and* **granita**.)

iced tea Tea, usually Indian, poured over ice and served cold garnished with slices of lemon or **mint** leaves.

Illawarra plum (*Austral.*) Fruit of *Podocarpus elatum*, also known as the brown pine plum. The fruit is small and dark blue and is usually made into **preserve**. (From Illawarra district, SE NSW.)

Indian tonic water *See* **tonic**.

infusion (1) The immersion of vegetables, herbs, **tea** leaves or **coffee** in boiling liquid, wine or oils until the flavours have been absorbed, so producing an aromatic fluid.
(2) Method used to give **liqueurs** their flavour. (*See* Chapter 14.)
(3) Beverage or oil given its flavour by infusion.

insalata (*Ital.*) [insal-ahta] Salad, e.g. *insalata mista* (*mista* means 'mixed'.)

Irish coffee Liqueur coffee made using Irish whiskey. (*See* Chapters 14 *and* 15.)

Italian dressing Salad dressing of oil, vinegar, garlic and seasonings. (*Compare* French dressing.)

jaffle Toasted sandwich with sealed edges. (*Compare* waffle.)

jaggery (*Anglo-Ind.*) Dark, crumbly sugar made from sugar-cane or the sugar palm and tasting of treacle. It is used in Indian cuisine. (*See also* palm sugar.)

jalapeño (*Mex.*) [halla-pen-nyoh] Variety of large, very hot chilli (*Capsicum annuum*). Dried smoked jalapeños are called chipotles.

jambon (*Fr.*) (masc.) [zjom-bo(n)] Ham.

jardinière, à la (*Fr.*) [ah lah jar-danee-air] Garnished with vegetables, esp. carrots, turnips, French beans, cauliflower and peas. (A *jardinière* is a 'female gardener'.)

jasmine tea China tea scented with jasmine flowers.

jeroboam (jerruh-boh-am) Large bottle of sparkling wine (e.g. champagne) containing the equivalent of 4 standard (750mL) bottles. (Jeroboam was King of Israel, 931–910 BC.) (*See also* champagne bottle sizes.)

Jerusalem artichoke *See* artichoke.

jugged hare (*UK*) Jointed hare casseroled in its own blood with wine, vegetables and seasonings.

julienne (*Fr.*) (fem.) [zhoo-lee-en] (1) Vegetables or meat cut into thin, evenly sized strips for preparation in cooking, or for use as a garnish, esp. for consommés. (Perhaps named after a French chef called Jean Julien. The word 'julienne' has been used in cooking since the early 18th century.) (2) Clear vegetable soup.

jus (*Fr.*) (masc.) [zjoo] Meat juice, or unthickened pan juices, gravy, stock. (*See* jus, au *and* jus-lié, *below.*)

jus, au (*Fr.*) [oh zjoo] (Meat) dressed with its own juice.

jus-lié (*Fr.*) [zjoo lee-ay] Thickened meat juice, gravy. (Lié means 'bound'.) (*See also* jus *above.*)

kabana/cabanossi (*Pol.*) [ka-bahna] Long, thin smoked sausage made from pork and beef seasoned with herbs and served cold.

kaffir lime Lime with knobbly skin. The fragrant leaves are used to flavour curries and infuse liquids, esp. Thai soups, with a citrus flavour. *Also known as* makrut lime.

Kahlua Proprietary Mexican liqueur based on cane spirit flavoured with coffee.

Kalamata olive Large black olive from Greece. (Kalamata is a town in S Greece.) (*See also* Ligurian olive *and* olive.)

kartoffel (*Ger.*) [car-toff-ell] Potato.

kasaundi/kasoundi (*Ind.*) [ka-sound-ee] Spicy tomato and onion relish.

kassler/kasseler (*Ger.*) [kass-lair] Pickled, smoked pork cutlets thinly sliced and served cold, or carved in thicker pieces and fried. (Named after a butcher from Berlin called Kasel.)

kebab/kabab/kebob (*Turk.* & *Ind.*) Small pieces of meat or fish, with vegetables, cooked on a skewer. (*See also* **brochette, döner kebab** *and* **shish kebab.**)

kedgeree (*Anglo-Ind.*) Dish of cooked rice, **smoked** fish, hard-boiled egg and parsley, flavoured with **turmeric** or **saffron**, usually served for breakfast.

ketchup Sauce made from **tomatoes**, mushrooms, **vinegar** and spices and used as a **condiment**. (In USA known as 'catsup'.)

kibbeh (*Arab.*) General term for the many dishes made from **burghul** used in the Middle East, esp. Lebanon.

kilojoule Metric measure of the energy in food (abbreviation kJ.)

Kilpatrick (*Austral.*) Sauce, the essential ingredients of which are bacon, **tomato sauce** and **Worcestershire sauce**, typically served hot (baked) on oysters as 'oysters Kilpatrick'.

kimchee (*Korea*) [kim-chee] Hot **pickled** cabbage served with almost all Korean meals.

King Island Island in Bass Strait noted for its dairy produce, e.g. King Island **cream, crème fraîche, cheddar,** and **cream cheeses** such as **brie** and **camembert**. Meats, e.g. herbed **salami** and bushman's bacon, are also prepared there.

kipfler potato Variety of small potato with waxy consistency ideal for salads.

kirsch [keersh] Style of **cherry brandy**. (*Kirsche* is German for 'cherry'.)

knackwurst (*Ger.*) [nack-voorst] Coarse-textured sausage traditionally served with **sauerkraut**.

Kobe beef [kohbay] Fine grain-fed and milk-fattened beef prized in Japanese cuisine. (Kobe is a city near Osaka, S Honshu.)

kohlrabi [coal rah-bee] Turnip-like vegetable with a swollen, edible stem.

korma (*Ind.*) [kor-mah] Mildly spiced casserole of meat or chicken cooked in a rich sauce usually enriched with ground coconut, cashew or other nut.

kosher (*Heb.*) [koh-sher] Lawful for Jews; meat killed or food prepared according to Jewish law. (*See also* **halal**.)

kugelhopf (*Ger.*) [koo-gel-hohpf] Yeast cake containing dried fruit. It is baked in a kugelhopf mould (a cake tin with a central funnel) so that it forms a crown-like shape.

kumara (*NZ*) [ku-mahra] Sweet potato; yam.

kümmel [kewmel] Dutch **liqueur** flavoured with caraway seeds, usually served on ice as a **digestif**.

la (*Fr.*) (fem.) [lah] The, when used with a feminine noun, e.g. *la* **maison**. (*See also* **le** *and* **les**.)

ladyfinger Finger-shaped sponge cake. (In French *langues de chat*, literally 'cat's tongues'.)

lady fingers/lady's fingers (1) **Okra** *or* **gumbo**.
(2) Small stubby bananas.

lager (*Ger.*) [lahg-er] Bottom-fermented **beer**. Most Australian beers are lagers. (*See* Chapter 14.)

lait, au (*Fr.*) [oh lay] With milk, e.g. **café au lait**. (*Lait* means 'milk'.)

laksa (*Singa.*) Spicy **seafood noodle** soup flavoured with coconut. (*See also* **tom yam kung**.)

langouste (*Fr.*) (fem.) [lahn-goost] Salt-water **crayfish** or **rock lobster**. (*See also* **lobster** *and* **homard**.)

langoustine(s) (*Fr.*) (fem.) [lahn-goose-teen] **Scampi**.

langue (*Fr.*) (fem.) [lahn-guh] Tongue.

lapin (*Fr.*) (masc.) [lapi(n)] Rabbit.

lardon (*Fr.*) (masc.) [lardo(n)] Strip of pork fat or bacon inserted into raw meat to enhance flavour while cooking, or cooked and served as a garnish with salads and vegetables.

lasagne (*Ital.*) (*pl.*) [laz-an-yuh] Large flat pieces of **pasta**; layered dish of lasagne sheets alternating with layers of meat-and-tomato sauce and cheese sauce.

lassi (*Ind.*) [lussi] Refreshing drink made with blended yoghurt, water and fruit. (*Compare* **smoothie**.)

latte, caffè *See* Chapter 16.

lavash (*Iran*) Soft flatbread; very thin, soft unleavened bread. Also known as **mountain bread**. (*Compare* **pita**.)

lavoche (*Yid.*) Crispbread; cracker.

layer (verb) Cocktail-mixing term. To pour the ingredients into the glass so that they don't mix but remain one above the other in clear layers.

lazy Susan *See* **dumb waiter** (2).

le (*Fr.*) (masc.) [luh] The, when used with a masculine noun, e.g. *le* jus. (*See also* **la** *and* **les**.)

legume Vegetable with seeds in a pod, e.g. bean, **lentil** and pea. (*See also* **légumes** *below*.)

légumes (*Fr.*) (masc. pl.) [lay-goom] Vegetables. (*See also* **classic menu** *and* **legume** *above*.)

lemon grass Tough stalk of a grass-like herb used to flavour food, esp. in Thai cooking. Only the inner bottom core is used. (In French, *citronelle*.)

lentil Dried bean (*Lens esculenta*) rich in nutrients including **protein**, one of the earliest cultivated foods known to humans; the basic ingredient of **dahl**. (*See also* **Puy lentils** *and* **split pea**.)

les (*Fr.*) (*pl.*) [lay] The, when used with a plural noun, e.g. *les* **légumes**. (*See also* **la** *and* **le**.)

liaison (*Fr.*) (fem.) [lee-ay-zo(n)] Mixture, typically egg yolks and **cream**, or flour and butter, used to thicken sauces and soups.

lié (*Fr.*) [leeay] Thickened (as in **jus** lié). (*Lier* means 'to bind', 'to thicken'.)

lifestyle diet A preference to eat or reject certain foods for various reasons.

Ligurian olive Tiny black Italian olive. (Liguria is a province in NW Italy.) (*See also* **Kalamata olive** *and* **olive**.)

lilly-pilly/lillipilli (*Austral.*) Tree (genus *Syzgium*) whose berries are small and crisp. They range in colour from white through pink to purple and are used to flavour desserts or to make jam or flavoured vinegar. The magenta lilly-pilly (*Syzgium paniculatum*) is the most common in SE Australia; its flavour is sweetly spicy.

linguini (*Ital.*) (*pl.*) [lingweenee] Type of **pasta** cut into long, very thin strips with square-cut edges.

liqueur [li-cure] Spirit- or wine-based **liquors** sweetened and flavoured with aromatic substances; often served as **digestifs**, e.g. **Grand Marnier**.

liqueur coffee Coffee served with a **nip** of **spirit** or **liqueur**. (*See* Chapters 14 and 15.) (*Compare* **corretto**.)

liquor [li-cur] Any liquid, but usually an alcoholic drink.

lobster (1) Large salt-water shellfish with big claws, genus *Homarus* (**homard** in French) with flesh and taste like that of **crayfish**. Real lobsters are not found in the waters of the southern hemisphere.
(2) (*Austral.* & *NZ* only) Southern **rock lobster** commonly known as **crayfish** or **cray**. (*See also* **langouste**.)

lobster thermidor *See* **thermidor**.

loin Joint of meat that includes some or all of the ribs. (*See also* **carré**, **rack** *and* **saddle**.)

Lorraine *See* **quiche Lorraine**.

lumpfish roe Eggs of the lumpfish, often referred to as **caviare** (which is actually the superior sturgeon's roe). Lumpfish roe is dyed red or black in imitation of real caviare.

lungo (*Ital.*) Style of espresso coffee—a 'long black'.

lyonnaise, à la (*Fr.*) [ah lah lee-on-nayze] In the style of Lyons (large city in SE France); garnished with onions and, often, potatoes. (*See also below*.)

lyonnaise, sauce (*Fr.*) [sohs lee-on-nayze] **Reduction** of white wine, **vinegar** and onions added to a **demi-glace**. (*See also* **lyonnaise, à la**, *above*.)

macadamia (*Austral.*) Nut from rainforest tree of the genus *Macadamia* used in both sweet and savoury dishes. Also known as bopple nut or Queensland nut. (J. McAdam, *d.* 1865, was secretary of the Philosophical Institute of Victoria.) (*See also* **macadamia oil** *below*.)

macadamia oil Oil of the **macadamia** nut.

macaroon Small almond-flavoured cake or biscuit.

macchiato [makee-ahto] *See* Chapter 16.

macédoine (*Fr.*) (fem.) [mass-ay-dwaan] **Diced** fruit or vegetables; fruit thus cut and set in jelly; fruit salad. (*Macédoine* is French for 'Macedonia'.)

macerate [mass-errate] To steep in liquid; to soak food, particularly fruit, in alcohol. (*See also* **marinate**.)

madeira **Fortified** wine with a distinctive 'burnt' taste (from the Portuguese island of Madeira). (French, *madère*.)

madeleine (*Fr.*) (fem.) [madel-ayn] Small sponge cake shaped like a **scallop** shell.

magnum Large bottle of **sparkling wine** (e.g. **champagne**) containing the equivalent of 2 standard (750mL) bottles. (*Magnum* means 'big thing' in Latin.) (*See also* **champagne bottle sizes**.)

maison/à la maison/de maison (*Fr.*) (fem.) [ah lah/duh may-zo(n)] Home-made; made to the **establishment**'s own special recipe, e.g. **pâté** *maison*. (*Maison* means 'house'.)

maître d'hôtel (*Fr.*) [met-ruh doe-tel] Steward, head-waiter. (*Maître* means 'master'.) (*See also* **beurre maître d'hôtel** *and* **maître d'hôtel, sauce**, *below*.)

maître d'hôtel, sauce (*Fr.*) (fem.) [sohs metruh dohtel] Sauce of melted butter, parsley and lemon juice. (*See* *also* beurre maître d'hôtel *and* maître d'hôtel.)

maize/sweet corn *See* corn.

malbec [mahlbek] Black grape variety popular in blends, e.g. cabernet-malbec, because it softens the effect of the tannin.

malt Grain, esp. barley, steeped in water to germinate and then dried and used for brewing beer and to make malt whisky. The process converts starch into maltose. (*See also* malt extract *and* malt vinegar *below*.)

malt extract Thick dark brown syrup used as an alternative to sugar or golden syrup.

malt vinegar Strong, dark brown vinegar distilled from malt and used mostly as a pickling agent; traditionally sprinkled over fish and chips. (*See also* balsamic vinegar, rice vinegar, vinaigrette *and* wine vinegar.)

malt whisky Traditional kind of Scotch whisky made from malted barley and distilled in a pot still.

mange-tout (*Fr.*) (masc.) [monje too] Thin green pea, eaten whole, pod and all. Also called snow pea and sugar pea. (Literally 'eat everything'.)

maraschino [maras-keenoh] Italian liqueur based on distillation of sour (Maraska) cherries.

maraschino cherries Cherries, dyed bright red, preserved in a type of maraschino juice.

marbling The streaks of fat contrasting with the lean in meat.

Margarita Cocktail made with tequila, cointreau and lemon juice.

margherita (*Ital.*) Simple style of pizza topping containing tomato, cheese and herbs.

marinade (noun) Mixture of liquids (e.g. wine, lemon juice, vinegar, oil, soy) with herbs and seasonings, in which food is marinated to enhance its flavour.

marinara (*Ital.*) [marin-ahra] With seafood, e.g. spaghetti *marinara*. (*Il mare* means 'the sea'.)

marinate/marinade (verb) To soak food in a marinade. (*See also* macerate.)

marinière, à la (*Fr.*) [ah lah marin-y-air] Method of preparing shellfish or fish in white wine with onions etc., e.g. moules *marinières*.

marjoram Compact shrub with small grey leaves and white flowers, both of which are aromatic and used as a seasoning herb. Similar to, but more subtle than, oregano, it is one of the ingredients of a traditional bouquet garni.

marmalade A preserve or jam usually made from bitter oranges and customarily served with toast at breakfast.

Marmite (*UK*) [mah-mite] Proprietary name for a yeast extract similar to Vegemite.

marmite (*Fr.*) (fem.) [mar-meet] Cooking vessel, esp. one with two handles in which soup is cooked.

marmite petite (*Fr.*) [mar-meet pu-teet] (1) Clear soup or broth. (*See also* consommé.) (2) Small cooking pot. (*Petite* means 'small'. *See also* marmite *above*.)

marron (1) (*Austral.*) [ma-ron] Freshwater crayfish (*Cherax*

tenuimanus) from WA, larger than a **yabby**.
(2) (*Fr.*) (*masc.*) [ma-rro(n)] Chestnut (as in *marrons* **glacés**, 'candied chestnuts').

marrow (1) Soft gelatinous substance found in hollow bones.
(2) White-fleshed gourd resembling a giant **zucchini**.

marsala [mar-sah-lah] Style of fortified red (almost brown) sweet **dessert** wine. Not to be confused with the Indian **masala** (spice).

marshmallow A soft sweet made of sugar, egg white, **gelatine**, etc.

Martini (1) Brand of **vermouth**.
(2) **Dry Martini** (cocktail).

maryland (*Austral.*) Whole thigh and leg joint of chicken. (*See also* **Maryland, chicken** *below*.)

Maryland, chicken (*USA*) Chicken portions, coated in breadcrumbs, deep-fried. (Maryland is a state in E USA.) (*See also* **maryland** *above*.)

Maryland cookies Chocolate-chip biscuits. (Maryland is a state in E USA.)

marzipan Thick paste of ground almonds and egg white, used in confectionery, esp. **petits fours**, and to hold decorative icing on fruit cake.

masala (*Ind.*) [ma-sah-lah] Aromatic spice, e.g. **garam masala** ('hot spice'). Not to be confused with **marsala** (wine).

mascarpone (*Ital.*) [mass-kur-ponay] Thick, rich cheese used in place of **cream** in some sweet dishes. (*See also* **crème fraîche** *and* **ricotta**.)

mash Mashed vegetable, usually potato, crushed and softened with butter, milk or cream. Other vegetables, e.g. parsnip, can be mashed in which case they are referred to as parsnip mash, **celeriac** mash, etc.

mask To coat food thoroughly in a sauce before serving. (*See* **nappé**.)

matsutake mushroom *See* **pine mushroom**.

matzos/matzoth (*Yid.*) (*pl.*) Wafers of unleavened bread served at the Passover.

mayonnaise Dressing or sauce of egg yolks and oil, flavoured with **vinegar**, salt, **pepper** and **mustard**.

MC Master of Ceremonies (at formal banquets, etc.).

medallion [med-al-ee-on] Small round piece of meat shaped like a medallion and usually skinned and boned. (In French, *médaillon* [may-dye-yee-o(n)].)

Mediterranean vegetables Typically a mixture of **capsicum**, eggplant, tomatoes and onion with herbs and sometimes olives.

mélange (*Fr.*) (*masc.*) [may-lahnzj] Mixture; medley.

melanzane alla parmigiana (*Ital.*) [melan-zahnay alla parmy-yahna] Baked dish of layered **eggplant**, tomato sauce, **mozzarella** and **parmigiano**. (*Melanzana* is Italian for 'eggplant'.)

Melba toast (1) Very thin toast made by splitting toasted bread and grilling the untoasted sides.
(2) Small thick squares of toast available commercially for use in **canapés**. (Named after Dame Nellie

Melba, 1861–1931, the famous Australian opera singer.) (*See also* **peach Melba**.)

menu (1) The range of food items served in an **establishment**.
(2) The arrangement by which the items are offered (set menu, à la **carte** menu, etc.).
(3) Written list of the food items served (also called the **bill of fare**). (*See* Chapters 2 *and* 6; *see also* **classic menu, menu d'agrément, menu dégustation, prix fixe** *and* **table d'hôte**.)

menu d'agrément (*Fr.*) (masc.) [menu dag-raymo(n)] Fixed price menu. (*See also* **prix fixe** *and* **table d'hôte**.)

menu dégustation (*Fr.*) Small portions of a range of the specialities of the **establishment**. (*Dégustation* means a 'sampling' or 'tasting'; Italian *degustazione*.) (*See* Chapter 2.)

merguez (*N Africa*) [mair-gway] Spiced sausage made from beef (not pork, since it is forbidden to Muslims), flavoured with **harissa**.

meringue [me-rang] Egg white whipped with sugar and baked. (*See also* **pavlova**.)

merlot [mairloh] Black grape variety noted for its **bouquet** and popular in blends, esp. **cabernet**-merlot.

mesclun Mixed salad leaves (esp. curly **endive**) of varying colours and flavours.

méthode champenoise [may-toad sham-pen-warz] Method by which French champagne is made. (*See* Chapter 14.)

methuselah (meth-ooz-uh-lah) Large bottle of **sparkling wine** containing the equivalent of 8 standard (750mL)

bottles. (Methuselah is the oldest man in the Bible, said to have lived to be 969.) (*See also* **champagne bottle sizes**.)

meunière, à la (*Fr.*) [ah lah mur-nee-air] Served with a sauce of browned butter, lemon juice, parsley and seasoning. (*Meunière* means 'miller's wife' or 'female miller'.)

mezcal Mexican spirit distilled from the *mezcal azul* cactus. **Tequila** is a superior type of mezcal.

meze/mezze (*pl.* **mezedes**) (*Grk* & *Turk.*) [mez-zay/meze-thes] Substantial assortment of cold **hors d'oeuvres** including such items as **taramasalata** and **dolmades**, often almost a meal in itself. (Literally 'half'—a half meal, the **hors d'oeuvre**). (*See also* **antipasto**.)

mignon (*Fr.*) [mee-nyo(n)] Small and dainty, as in **filet mignon**.

mignonette (*Fr.*) (fem.) [mee-neeyon-ett]
(1) Coarsely ground white **pepper**.
(2) Small cuts of meat elaborately prepared, e.g. **mignonnettes d'agneau**.
(3) Potatoes sliced into thin sticks.
(4) Variety of lettuce.

milanese (*Ital.*) [milan-ay-zay] In the style of Milan (N Italian city), e.g. **risotto alla milanese. mille-feuille** (*Fr.*) (fem.) [meel fuh-ee] Puff-pastry cake layered with whipped **cream** and jam. Some savoury *mille-feuilles* are layered with a meat or fish sauce. (Literally 'thousand-leaf'.)

minced meat/mince Any meat ground or minced. Not to be confused with **mincemeat**, *below*.

mincemeat (*UK*) Fruit mince; sweet spicy mixture of currants, apple, etc.

usually with **suet** and **brandy**, used in mince pies at Christmas. Not to be confused with **minced meat**, *above.*

mineral water Water flavoured with minerals, often thought to be health-promoting. Sparkling mineral water is aerated with carbon dioxide, either 'naturally' (at its source) or by injection. (*See also* Chapter 14.)

minestra (*Ital.*) [min-ess-tra] Soup. (*See also* **brodo**, **minestrone** *and* **zuppa**.)

minestrone (*Ital.*) [min-ess-troh-nay] Thick vegetable and **pasta** soup. (*See also* **minestra**.)

mint Herb, of which there are many varieties (genus *Mentha*), used as a flavouring or **garnish** in sweet or savoury dishes. (*See* **crème de menthe**.)

mint sauce Thin sauce, served cold, of finely chopped garden **mint**, **malt vinegar** and sugar, served traditionally as an **accompaniment** to roast lamb.

mint, Vietnamese *Polygon um oloratum*, also known as 'hot **mint**'; used as **garnish** in Vietnamese **cuisine**.

mints, after-dinner Chocolates with **mint**-flavoured fondant centres served with **coffee** after the meal.

minute steak [minit] Tender steak cooked for one minute only. (It is therefore not necessary to ask how the guest prefers the steak cooked.) (*See also* **entrecôte**, **filet mignon**, **porterhouse**, **rump**, **sirloin** *and* **T-bone**.)

mirepoix (*Fr.*) [meer-pwah] **Diced** carrot, onion and celery **sautéed** and used to flavour soups and stews. *Mirepoix au gras* is this mixture cooked with ham or bacon. (Created by chef to the Duc de Mirepoix, 1699–1757, French general.) (*Compare* **soffritto**.)

mirin (*Jap.*) Sweet rice wine, low in alcohol, used only for cooking in Japan. (*See also* **sake**.)

miroir (*Fr.*) (masc.) [meer-wah] Shiny fruit **glaze** completely covering a dessert. (Literally 'mirror'.)

mise-en-place (*Fr.*) (fem.) [meez-o(n)-pluss] The setting out of equipment, garnishes, etc. in preparation for service or cooking. (Literally 'set in place'.) (*See* Chapters 4 *and* 13.)

miso (*Jap.*) [meessoh] Paste made from fermented **tofu** and used to make **miso shiru**.

miso shiru (*Jap.*) Soup made from **miso** and **stock**, often eaten at breakfast time.

misto/mista (*Ital.*) (adj.) Mixed, e.g. **fritto misto, insalata mista**.

mL Millilitres.

mocha/mocca [mokkah] (1) Strong variety of **coffee** bean.
(2) A combination of **coffee** and chocolate flavouring (as in mocha **coupe**).
(3) Coffee drink. (*See* Chapter 16.) (From Mocha, or Al Mukha, port in Yemen, from which the coffee beans were shipped.)

mochaccino (*Austral.*) Style of coffee—a short espresso mixed with hot chocolate in a cup and topped with bitter chocolate and **marshmallows**.

mocktail Non-alcoholic, or **virgin**, **cocktail**.

môde, à la (*Fr.*) [ah lah mode] In the manner of, e.g. *à la môde de Paris*. (*See also* **boeuf** à la **môde**.)

mole (*Mex.*) [molay] **Sauce** made with a variety of **chillies** and spices and any of many other ingredients—nuts, seeds, dried fruit, etc. It often contains bitter chocolate, e.g. the traditional *mole poblano* (chilli turkey stew).

mollusc(s) Type of shellfish having a soft unsegmented body (unlike **crustaceans**) and usually a shell. Molluscs include gastropods (snails), bivalves (**oysters, scallops**, mussels) and cephalopods (octopus, cuttlefish, squid).

monosodium glutamate *See* **MSG.**

morel [morrell] Variety of mushroom (*Morchella esculenta*) with a crinkled and brownish pointed cap resembling the pitted bark of a tree. (*See also* **cep, oyster mushroom, pine mushroom, shitake** *and* **truffle.**) (Not to be confused with **morello.**)

morello [mur-elloh] Dark, bitter cherry commonly used in jams and cooked **puddings.** (Not to be confused with **morel.**)

Moreton Bay bug *See* **bug.**

mornay Cheese sauce made with milk or **cream** to which dried **mustard** has been added. (Philippe de Mornay, 1549–1623, colleague of Henri IV of France.) (*See also* **gratin, gratinée.**)

mortadella (*Ital.*) [mort-a-dell-ah] Lightly **smoked** large Italian sausage, pale in colour with spots of pork fat. It is sometimes flavoured with **parsley, pistachios, peppercorns, coriander** seeds or green **olives** and served with the **antipasto.** (Mortadella is not a place; the origin of the name is disputed.)

moselle (1) Rather sweet white wine from the Mösel River region of Germany.

(2) **Generic** wine blended to resemble the style of German moselle.

moule (*Fr.*) (fem.) [mool] Mussel, e.g. *moules* marinières.

mountain bread Thin, flat, unleavened bread. (*See also* **lavash** *and* **pita.**)

moussaka (*Grk*) [moose-ah-kah] Minced meat and eggplant baked with a topping of cheese sauce.

mousse (*Fr.*) (fem.) [moose] (1) Light and frothy dish, usually set with **gelatine,** in which the basic ingredient is folded with whipped **cream.** It can be sweet or savoury (*mousse de saumon,* 'salmon mousse'; *mousse au chocolat,* 'chocolate mousse').
(2) The bubbles in **sparkling wine.**

(*Mousse* means 'foam', 'lather' or 'effervescence'.)

mousseline (*Fr.*) (fem.) [moose-leen] (1) **Puréed** fish or meat blended with **cream.**
(2) Sauce to which **cream** has been added, e.g. **hollandaise** *mousseline.*

moutarde (*Fr.*) (masc.) [moo-tard] Mustard.

mozzarella (*Ital.*) [motsa-rella] Italian cheese, waxy in texture, which when cooked becomes stringy; the classic topping for pizza. (*See also* **bocconcini, cheddar, parmesan** *and* **ricotta.**)

MSG Monosodium glutamate. Salt used to enhance the flavour of meat, much used in Chinese cooking and as an additive in processed foods. Some people object to MSG being included in their food since it is thought it may cause an allergic reaction.

muddle Cocktail-mixing term. To crush in the bottom of the glass with a special 'muddler' (miniature pestle) or the base of a bar spoon. (*See* **Old Fashioned.**)

muffin (1) (*UK*) Round, flat yeast bun split and toasted and served buttered.

(2) (*USA*) Large cup cake usually containing fruit, e.g. blueberry muffin.

mulled wine Red wine heated gently with sugar, spices, raisins and **citrus** fruits and served hot. (*See also* **glühwein.**)

mulligatawny (*Anglo-Ind.*) [mully-guh-tawny] Highly seasoned chicken soup with vegetables. (Literally 'pepper water' in Tamil.)

muscat (1) Australian style of sweet fortified **dessert wine**, e.g. Rutherglen muscat.
(2) One of a number of varieties of grape, also known as **frontignac.**

muscatel/muscadel   [muss-kuh-tel] Raisin; dried fruit of **muscat** grape.

muselet (*Fr.*) [moo-sell-ay] Wire muzzle used to hold in the corks of **sparkling wines.** (*See* Chapter 15.)

mushroom One of various kinds of edible fungus but usually the cultivated mushroom (*Agaricus bisporus*). (French *champignon*.) (*See also* **cep, morel, oyster mushroom, pine mushroom, porcino, shitake** *and* **truffle.**)

mustard Yellow or brown **condiment** made from seeds of the mustard plant (genus *Brassica*). There are many different ready-made mustards, including hot English mustard, the milder Australian and German mustards, sweet American mustard,

and various French mustards, including some containing mustard seeds. Mustard is also available in powdered form. (French *moutarde*.) (*See also* **cress** *and* **Dijon.**)

mustard fruits Relish of dried fruit, esp. apricots, figs or pears, **pickled** in a mixture of wine, honey and mustard (Italian *mostarda di frutta*).

naan/nan (*Ind.*) [nahn] Flat, puffy, leavened bread cooked in a **tandoor.**

nacho (*Mex.*) [nah-choh] Corn chips served with spicy sauce of beans, tomatoes and/or meat topped with grated cheese, soured **cream** or **guacamole.** (*See also* **enchilada, taco** *and* **tortilla.**)

nam pla (*Thai.*) [nam plah] Salty, pungent fish sauce used in cooking. (*See also* **nuoc mam.**)

nam prik (*Thai.*) [nam prik] Spicy shrimp sauce served as a **condiment.**

napkin *See* **table napkin.**

Napoléon (brandy) Marketing term without technical or legal meaning used on labels of brandy, particularly **cognac.** (Napoléon Bonaparte, 1769–1821, was Emperor of the French.)

Napoli (*Ital.*) From Naples. (*See* **napolitana** *below.*)

napolitana (*Ital.*) [napol-it-ahna] From Naples, town in N Italy. Pizza *napolitana* contains **tomato,** cheese, herbs, **garlic, olives** and **anchovies.**

nappé(e) (*Fr.*) [nap-pay] Covered evenly with sauce or jelly. (*See* **mask.**)

nashi [naashi] Crisp, juicy fruit related to a pear but resembling an apple. There are two varieties: *chojuro* is a russet colour and *shinsui* is smooth and yellowy-green.

nasi (*Malay*) [nahsee] Cooked rice.

nasi goreng (*Malay*) [nahsee gore-eng] Fried rice with chicken or prawns, shredded **omelette**, onions and spices. (*Goreng* means 'fried'.)

native peach *See* **quandong.**

natural Uncooked, as in oysters natural. (*See also* **naturel, au.**)

nature (*Fr.*) (adj.) [natoor] Plain, unadulterated or without **garnish**; cooked with no sauce. (*See also* **naturel, au** *below*.)

naturel, au (*Fr.*) [oh na-toor-el] In a state of nature; food cooked plainly and simply, or not at all, e.g. **huîtres** *au naturel* (oysters **natural**).

navarin (*Fr.*) (masc.) [nav-a-ri(n)] Mutton (lamb) stew with vegetables and, originally, turnips. (*Navet* means 'turnip'.)

navy bean *See* **haricot.**

neat **Liquor**-mixing term; undiluted (e.g. neat **whisky**).

nebuchadnezzar Largest recognised size of bottle for **sparkling wine**, the equivalent of 20 standard 750mL bottles. (Nebuchadnezzar was the King of Babylon who conquered Judah and carried the Jews off into exile in 586 BC.) (*See also* **champagne bottle sizes.**)

newburg Method of cooking and serving shellfish, esp. **lobster** or **crayfish**, with a sauce of sherry, cream and egg yolk. (Originally created by a New York chef called Wenburg, the first

three letters of his name being transposed in the name of the dish.)

niçois(e) (*Fr.*) (fem. adj.) [neece-swaz] Literally 'from Nice'. (Nice is a city in SE France). Typically **provençal** dishes, containing **garlic**, **anchovies**, **olives**, green beans and **tomatoes**. The most famous dishes of this type are **salade niçoise** and **poulet sauté** *à la niçoise*.

nip (1) Small measure (of **spirits**); the standard Australian measure or nip is 30mL.
(2) (*UK*) Quarter bottle.

noble rot *See* **botrytis.** (*See also* Chapter 14.)

noisette (*Fr.*) (fem.) [nwah-zet] (1) Hazelnut; coloured or shaped like a hazelnut. (*See* **beurre noisette** *and* **pommes noisettes.**)
(2) Small round tender piece of meat, usually lamb, cut from the **loin** or rib.

noodle(s) Dough similar to **pasta**, usually cut into thin strands or rolled into thin sheets (used as wrappers in Chinese cooking). Most European noodles are made from wheat, flour and eggs, but some are made of rice and a few of potatoes. Asian noodles are made from rice or wheat and seldom contain eggs. (From the German *nudel*, meaning 'dumpling'.) (*See also* **hokkien**, **soba** *and* **udon noodle.**)

nori (*Jap.*) [no-ree] Kind of seaweed, dried and pressed into thin sheets, used as an edible wrapping or casing for **sushi.**

normande, à la (*Fr.*) [ah lah nor-mahnd] In the Normandy style; with apples and possibly **calvados** as well as cream, e.g. **pommes de terre** *à la* *normande.* (Normandy is a province in N France.) (*See also* **normande sauce** *below.*)

normande sauce Rich sauce made with **cream** and fish **stock** (French, **sauce** *normande*). (*See also* **normande, à la,** *above.*)

nose Wine-tasting term; the combination of the **aroma** and **bouquet** of a wine.

nougat [noo-gah] Soft, white, slightly chewy sweet made of sugar or honey, nuts, candied fruits and egg white. It is available in blocks that are cut into bite-sized pieces.

nuoc cham (*Viet.*) [nuok chum] **Nuoc mam** with **chilli**, **garlic**, sugar and lime juice added and used as a dipping sauce. (*Nuoc* means 'water' or iuice', *cham* 'dip'.)

nuoc mam (*Viet.*) [nuok mum] Salty, pungent fish sauce used in cooking. (*Nuoc* means 'water' or 'juice', *mam* 'sauce'.) (*See also* **nam pla.**)

nutrient Providing nourishment; a nourishing ingredient.

NV Non-vintage. Term particularly used for some **sparkling wines** to indicate that they are blended wines, not made entirely from the **vintage** of a particular year.

oeuf (*Fr.*) (masc.) [erf] Egg.

oeuf en cocotte [erf ahn ko-kot] Baked egg. (*See* **cocotte.**)

oeuf mollet [erf moll-ay] Soft-boiled egg.

oeufs à la neige (*Fr.*) [erf ah lah nayzj] Floating islands. (Literally 'eggs like snow'.)

offal The cheaper edible parts of an animal, esp. the internal organs. (French *les* **abats.**)

off-premises catering Food prepared and brought to a venue ready to serve.

oie (*Fr.*) (fem.) [wah] Goose, e.g. **confit** *d'oie.*

okra Seedpod of *Hibiscus esculentis,* also called **lady's fingers** or **gumbo.** Popular vegetable in Middle Eastern and **creole** cooking.

Old Fashioned Cocktail of **rye whiskey** and soda poured over sugar flavoured with **Angostura Bitters.** The sugar is often **muddled** in the glass, so that an Old Fashioned glass (*see* Chapter 13 *and* subsequent diagram of glassware) has a specially thick base.

olive (1) Small fruit of the olive tree, either green (immature) or black (mature). Green olives are often stoned and stuffed with **anchovy,** or a sliver of **capsicum** or almond. Olives are included in many dishes, but are also served as an **appetiser,** esp. on an **antipasto** platter, or used as a garnish in cocktails. (*See also* **Kalamata, Ligurian** *and* **Spanish** olive, *and* **olive oil** *and* **tapenade.**) (2) **Paupiette** (beef olive).

olive oil Oil extracted from **olives.** 'Pure' olive oil is pressed under heat from the pulp and kernels of lower-grade olives. **Virgin** and extra-virgin oil are finer and simply crushed from the first pressing of the highest-grade fruit.

oloroso Golden and sweet style of Spanish **sherry.**

omelette Eggs beaten with seasonings and fried in hot butter until just set; fillings, e.g. mushrooms or cheese, are sometimes added before the omelette is turned from the pan.

on-the-rocks Cocktail-mixing term: poured over a large quantity of cubed ice (e.g. **Scotch** on-the-rocks).

oregano Leaves of plant similar to **marjoram** used as robust seasoning herb, esp. in Mediterranean cooking.

organic Label given to food, either meat or vegetables, produced on farms that use no artificial fertilisers. Farms certified as organic use methods of production that care for the soil and environment.

osso buco (*Ital.*) [oss-oh boo-koh] Rich **casserole** of knuckle of **veal** in a **tomato**, wine and vegetable sauce, traditionally served with rice garnished with **gremolata**. (*Osso* is Italian for 'bone' and *buco* means 'hole'—'bone with a hole' in it.)

ouzo/oyzo (*Grk*) Aniseed-flavoured **spirit**, similar to **pastis**.

oven-dried tomato *See* **tomato, dried**.

overproof Alcoholic drinks stronger than 57.1% alc/vol, e.g. overproof **rum**.

oyster Bivalve **mollusc** (shellfish) usually eaten raw. The variety most commonly served in Australia is the Sydney rock oyster (*Saccostrea commercialis*). The New Zealand bluff oyster (*Ostrea luteria*) has a distinctive flavour and is used in soups and seafood casseroles. (*See also* **huître**, **Kilpatrick** *and* **natural**.)

oyster mushroom Large white mushroom (*Pleurotus ostreatus*) with a slight taste of **oysters**. Also known as abalone mushroom. (*See also* **cep**, **morel**, **pine mushroom**, **shitake** *and* **truffle**.)

oyzo *See* **ouzo**.

paella (*Span.*) [pie-yella] Rice dish flavoured with **saffron** and containing chicken, **shellfish**, **chorizo**, peas and other vegetables.

paillard (*Ital.* from *Fr.*) [pie-yar] **Escalope** of **veal**. (Paillard was a 19th-century Parisian restaurateur.)

pain (*Fr*) (masc.) [pa(n)] Bread. (NB **Pane** in Italian. *See also* **pané**.)

pain au chocolat (*Fr.*) [pa(n) oh shokohlah] **Croissant** containing chocolate. (Literally 'chocolate bread'.)

pakora/pakhora/pakorha (*Ind.*) [pack-or-ah] Savoury **fritter** made of **chick-pea** flour containing spicy **diced** vegetables.

palate Upper part of the mouth; sense of taste.

palm sugar Coarse, dark sugar obtained from the sap of the palmyra palm. It is used in Malaysian and Indonesian savoury dishes. (*See also* **jaggery**.)

pancake Thin, flat, fried **batter** cake. (*See* **crêpe**.)

pancetta (*Ital.*) [panchetta] Pork belly rolled and eaten hot or cold; bacon.

pane (*Ital.*) [pan-nay] Bread. (NB *Pain* in French; *see also* **panino**.)

pané(e) (*Fr.*) [pan-ay] Fried in breadcrumbs. (*See also* **pain**.)

panettone (*Ital.*) [panuh-tone-ay] N Italian Christmas cake made from yeast and containing dried fruit, spices and nuts.

panforte (*Ital.*) [panfortay] Sweetmeat containing **glacé** fruit and nuts.

panino (*Ital.*) (*pl.* panini) [pan-neenoh/pan-neeni] Bread roll. (*See also* **pain** *and* **pane**.)

pannacotta (*Ital.*) Cold dessert made with cream set with **gelatine** so that it 'shivers' but is not rubbery. The cream is sometimes flavoured with fruit or **liqueur**. (Literally 'baked cream'.)

pantry Storeroom, esp. for crockery, cutlery, etc.

panzanella (*Ital.*) [pan-zan-ella] Salad made with dry bread, tomatoes, onion, etc. (*Compare* **fattoush**.)

papillote (*Fr.*) (fem.) [pap-ee-yot] (1) Buttered paper; **en papillote**, cooked in buttered paper. (Fish is often served in this way.)
(2) Small **frill** of paper used to garnish **cutlet** bones.

pappadam/poppadum (*Ind.*) [poppa-dum] Very light and thin crisp wafer made from **lentil** flour cooked in oil, served with **curry**.

paprika [pap-reeka] Red mild spice (*Capsicum annuum*) bought powdered and used both as a **garnish** and for its colour and flavour in cooked dishes, esp. Hungarian **goulash**.

parfait (*Fr.*) (masc.) [par-fay] (1) Creamy iced **pudding**.
(2) Layered fruit and ice cream dessert served in tall glasses.
(3) Smooth, savoury **pâté**, usually of meat, blended with **cream**.

Parma ham Fine, raw **smoked** Italian ham from Parma, city in N Italy. (Italian **prosciutto** *di Parma*.)

parmesan (*Ital.*) [parm-i-zan] Very hard dry cheese much used in Italian cooking. (Originally from Parma, city in N Italy.) (*See* **parmigiano** *below*.)

parmigiano (reggiano) (*Ital.*) [parmi-zjahnoh rezj-ee-ahnoh] **parmesan** cheese produced in Parma, N Italy. (It can only carry this label if produced in the region.) (*See also* **grana**.)

par stock The correct amount or quantity of a product kept in stock (sufficient to meet demand, but not too much) as decided by management.

pasta (*Ital.*) Staple food of Italy made of durum wheat flour (**semolina**) or potato, mixed with water and, usually, egg and often sold dried. There are many varieties of pasta. (*See* **cannelloni, fettuccine, gnocchi, lasagne, linguini, penne, ravioli, risoni, spaghetti, stracci, tagliatelle, tortellini** *and* **vermicelli**; *see also* **al dente, carbonara, farro, minestrone** *and* **noodles**.)

pastis (*Fr.*) (masc.) [pas-teece] (1) Strong alcoholic drink flavoured with aniseed, similar to **ouzo**, e.g. **Pernod**.
(2) Any of several pastries made in SW France, e.g. *pastis Béarnais*. (Béarn is a province in SW France.)

pasto (*Ital.*) A meal. (*See also* **antipasto**.)

pastrami (*USA/Yid.*) [pass-trahmee] Cured, spiced roast beef **silverside** rubbed with spices and black **peppercorns**. It is usually sliced and served in **rye** bread sandwiches with **dill pickles**.

pâte (*Fr.*) (fem.) [paht] Dough, pastry or **batter**. (Not to be confused with **pâté**, below.)

pâté (*Fr.*) (masc.) [paht-ay] Paste of **blended** meats or fish, flavoured with seasoning, herbs and, sometimes, alcohol. (Not to be confused with **pâte**, above.) (*See also* **parfait, rillettes** *and* **terrine**.)

pâte à choux (*Fr.*) (fem.) [paht ah shoo] **Choux** pastry.

pâte brisée (*Fr.*) (fem.) [paht breez-ay] Short-crust pastry.

pâté de foie gras (*Fr.*) (masc.) [pat-ay duh fwa gra] Pâté of force-fed goose or duck liver made by **blending** the fattened livers with minced pork and **truffles**. (*Gras* is French for 'fat'.)

pâté en croûte (*Fr.*) (masc.) [pat-ay ahn croot] Pâté baked in a pastry case. (*See* **croûte**.)

pâte feuilletée (*Fr.*) (fem.) [paht fur-yet-ay] Puff pastry. (*See also* **feuilleté**.)

pâté maison (*Fr.*) (masc.) [pat-ay mayzo(n)] Home-made **pâté**, made according to the chef's own recipe. (*See* **maison**.)

pâte sablée (*Fr.*) (fem.) [paht sab-lay] Sweetened **shortcrust pastry**. (*Un sablé* means 'a shortbread'.)

pâte sucrée (*Fr.*) (fem.) [paht soo-cray] Rich, sweetened **shortcrust pastry**. (*Sucré* means 'sugared'.)

patisserie (*Fr.*) (fem.) [pat-ees-er-ee] (1) Pastry cake.
(2) Cake shop; tea-room.

patty Small flat cake of meat (e.g. the filling of a hamburger); a little pie.

pauchouse *See* **pochouse**.

paupiette (*Fr.*) (fem.) [paw-pee-ett] Thin slice of meat rolled with a stuffing (*paupiettes de* **veau**), sometimes referred to as an **olive** (beef olives).

pavé (*Fr.*) [pavay] (masc.) (1) Sponge cake or dessert of biscuits sandwiched in layers with creamy filling.
(2) Savoury **mousse** set in **aspic** and garnished with **truffles**.
(*Pavé* means 'paving stone'.)

pavlova (*Austral. & NZ*) Soft **meringue** cake filled with **cream** and topped

with fruit. (Named for Anna Pavlova, Russian ballerina, 1882–1931.)

pax The number of people for a meal, accommodation, etc., e.g. '8 pax' means eight people. (The term originated in the airline industry and is short for 'passengers'.)

paysanne, à la (*Fr.*) (masc.) [ah lah pay-ee-zan] **Diced** vegetables used as **garnish**, e.g. **potage** *à la paysanne*; **braised** dishes cooked with vegetables, e.g. côtes *de* veau *à la paysanne*. (*Paysanne* means 'peasant woman'.)

peach Melba Poached peach halves enclosing **vanilla** ice cream coated with raspberry **coulis** (French, pêches *Melba*). The original peach Melba was devised by famous chef Auguste Escoffier (1846–1935) in honour of the Australian opera singer Dame Nellie Melba. (*See also* **Melba toast**.)

pecan [pekahn/peecan] Smooth brown nut from the Mississippi region, USA; hickory nut.

pêche (*Fr.*) (fem.) [paysh] Peach. (*See also* **peach Melba**.)

pecorino (*Ital.*) [peck-oreenoh] Salted, hard sheep's-milk cheese, esp. *pecorino Romano*.

Peking One of the five main styles of Chinese cuisine; Peking cuisine is light and delicate. (Peking, now called Beijing, is a city in N China.) (*See also* **Cantonese, Fukien, Honan** *and* **Szechwan**.)

Peking duck (*Chin.*) The crisp, crackling skin of mandarin duck eaten in a **pancake** with **spring onion**, cucumber and **hoi sin sauce**. The flesh of the duck is served separately.

penne (*Ital.*) (*pl.*) [penay] Short tubes of **pasta** cut at an acute angle. (Literally 'quills'.)

peperonata (*Ital.*) [pepperon-ahta] Stew of red and yellow **capsicums**, sometimes with added onions, tomatoes and garlic. It is served either hot (as a vegetable) or cold (e.g. on an **antipasto** plate). (*Compare* **peperoni** *and* **ratatouille**.)

peperoni (*Ital.*) (*pl.*) [pepper-ohny] Preserved **sweet peppers**. (*Peperoni* is Italian for 'capsicums'.) (Not to be confused with **pepperoni**; *compare* **peperonata**.)

pepper Condiment made from the dried berry-like fruit (**peppercorn**) of the pepper plant (*Piper nigrum*). (*See* **peppercorn**; *see also* **capsicum**, **cayenne** *and* **sweet pepper**.)

pepperberry (*Austral.*) Round black berry from Australian tree *Cryptocarya obovata*. When dried and ground it is used as an extra-strong seasoning agent. The leaves are also used as a milder seasoning. (*Also known as* Mountain Pepperberry *and* Mountain Pepperleaf.)

peppercorn Seed of the **pepper** plant picked and processed at one of three distinct stages:
(1) *Green* peppercorns picked while the seed is still soft and green; usually bottled in **brine** but occasionally available fresh. (*See* **poivre vert**.)
(2) *Black* peppercorns picked when the skin has turned yellow and the kernel has hardened. The berry is dried once the skin has been removed.
(3) Once the berry turns red, the outer coating is washed off and the kernel

dried for *white* **pepper**. This has more bite but less **aroma** than the black peppercorn.
Black peppercorns are usually freshly ground from a pepper mill. Ground white pepper is contained in a **cruet** on the table. (*See also* **poivre, au**.)

pepperoni (*Ital.*) [pepper-ohny] Hard, spicy beef-and-pork **salami**. (Not to be confused with **peperoni**.)

peppers *See* **capsicum**, **peperoni** *and* **sweet pepper**.

percolator A coffee-making utensil in which boiling water rises through a tube before falling on to ground coffee through which it seeps or percolates.

Pernod [pair-no] **Proprietary** brand of **pastis**.

Perrier Brand of French **mineral water**. (*See* Chapter 14.)

persillade (*Fr.*) (*fem.*) [pear-sill-ahd] Mixture of **garlic** and chopped **parsley**. (*Compare* **gremolata**.)

persillé (*Fr.*) [pear-sill-ay] Dishes seasoned with a **persillade**, e.g. **boeuf** *persillade*.

pesce (*Fr.*) (*masc.*) [pesskuh] Fish.

pesto (*Ital.*) [pest-oh] Paste made by blending **basil**, **parmesan** cheese, **pine nuts**, **garlic** and **olive oil**. Variants, also described as 'pesto', replace the basil and pine nuts with different herbs and nuts. (*See also* **pistou**.)

petit four (*Fr.*) (*masc.*) (*pl.* petits fours) [put-ee foor] Small fancy cake or biscuit. There are different varieties: dry and undecorated plain petits fours; miniature cakes, iced **marzipan**

sweets, **friandises** and sugar-coated fruits. Savoury petits fours are served with **apéritifs**. (*Petit* means 'little' and *four* (usually) 'oven', so *petits fours* 'little baked things', 'little cakes'.)

petits pois *See* **pois**.

pho (*Viet.*) [fo] (short 'o' as in hot) Soup.

phosphorus [fos-fore-uss] Non-metallic element found in all foods. Very high intakes can inhibit the absorption of calcium.

phyllo *See* **filo**.

piccalilli A yellow pickle of chopped vegetables, mustard and hot spices. (Origin uncertain, perhaps from pickle + chilli.)

pickle (1) (noun) **Condiment** of fruit or vegetables preserved in **vinegar** with sugar and spices, e.g. **dill pickle**, mango pickle.
(2) (verb) To **preserve** vegetables or fruit in a **pickle** solution, e.g. pickled onions. (*See also* **brine**, **cure** *and* **smoke**.)

pièce de résistance (*Fr.*) (fem.) [pee-yayss duh ray-zees-tahns] The main dish of a meal, substantial and usually rather grand. (Literally 'item of strength'.)

pilaff (*Turk.*) [peelaf] Middle Eastern dish of spiced rice with meat, chicken or fish, usually cooked together. (*See also* **pilau** *below*.)

pilau (*Ind.*) [pih-lauw] Indian version of **pilaff**, typically including chicken and/or mutton or goat's meat. (*See also* **biryani**.)

pilsner/pilsener (1) Pale hop-flavoured **lager** originally made in Pilsen, Bohemia (now Plzen, in the Czech Republic).

(2) Style of beer glass. (*See* Chapter 13 *and* subsequent diagram of glassware.)

Pimm's Properly 'Pimm's No. 1 Cup', an English **proprietary** gin-based flavoured liquor, usually served in summer mixed with lemonade or **ginger ale** and ice, and garnished with cucumber and **mint**.

pine mushroom Large, golden mushroom (*Armillaria edodes*) which grows under red pine trees. Also known as **matsutake mushroom**. (*See also* **cep**, **morel**, **oyster mushroom**, **shitake** *and* **truffle**.)

pine nut The seed of the Mediterranean stone pine. Pine nuts are an essential ingredient of **pesto**.

Pink Gin Cocktail of gin served with a dash of **Angostura Bitters**.

pinot noir [pee-no nwaar] Grape variety used in the famous French red **burgundy** wine. Pinot noir wines are usually soft and light; they are also part of the blend used to make French **champagne**.

pipe To squeeze **cream**, icing, etc. in decorative patterns on a cake or other food item.

piperade (*Fr.*) (fem.) [peep-ay-rahd] Stewed **tomato**, **capsicum**, onion and **garlic** mixed with beaten eggs and cooked slowly over gentle heat.

pipi Shellfish, either the smaller *Plebidonax deltoides* (Australia: about 5cm) or the *Amphidesma australe* from New Zealand which is larger (about 8cm wide). Pipis resemble small clams or cockles.

piquant [pee-kant] Spicy, appetising.

piroshki (*Russ.*) [pi-rohsh-kee] Savoury pasty.

Piscatorian A person who eats fish (all seafood), cheese, milk, cream, yoghurt, butter and eggs, but no meat or poultry.

pistachio [piss-tashee-oh] Fruit of the *Pistacia* tree, native to the Middle East. Pistachio nuts are used in savoury and sweet dishes, esp. cream desserts and **pilaus**, and with sliced meats where their bright green colour enhances the appearance of the food.

pistou (*Fr.*) (masc.) [pees-too] **Provençal** paste **condiment** made of sweet **basil** and olive oil with **garlic** and **parmesan** cheese, similar to Italian **pesto**. Pistou soup (**soupe** *au pistou*) contains a spoonful of pistou, which floats on top of the broth.

pita/pitta (*Grk*) [pitta] Flat round double-layered bread that can be cut open and filled; pocket bread. (*Compare* **lavash** *and* **mountain bread**.)

pithiviers (*Fr.*) [peeteev-ee-ay] **Puff-pastry** tart traditionally sandwiched with almond cream but often filled with a savoury mixture. The pastry top is usually **scalloped** around the edges and **scored** with curves radiating from the centre. (Pithiviers is near Orléans in central France.)

pizza (*Ital.*) [peetsa] Plate-like base of dough covered with topping that usually includes cheese, tomato and herbs. (*See* **capricciosa**, **margherita**, **marinara** *and* **napolitana**; *see also* **calzone** *and* **focaccia**.)

plat du jour (*Fr.*) (masc.) [pla doo zhjoor] Today's dish; dish of the day.

plate (1) A plate. (2) Silverware.

platter Large serving dish; a 'sizzling platter' is taken to the table direct from the stove and so is usually made of cast iron. (*See also* **skillet**.)

ploughman's lunch (*UK*) Cheese and **pickle** served with a hunk of bread and usually accompanied by beer.

plum pudding *See* **Christmas pudding**.

poché(e) (*Fr.*) (adj.) [poshay] Poached.

pochouse/pauchouse (*Fr.*) [poh-shooz] Fish stew from Burgundy cooked with white wine. (Dialect version of *pêcheuse*, 'fisherwoman'.) (*See also* **bouillabaisse**.)

point, à (*Fr.*) [ah pwa(n)] Medium; steak cooked so that it shows a tinge of pink blood. (*See also* **bien cuit, bleu** *and* **saignant**.)

poire (*Fr.*) (fem.) [pwaar] Pear.

poireau(x) (*Fr.*) (masc.) [pwah-roe] Leek(s).

pois (*Fr.*) (fem.) [pwah] Pea, usually eaten young as *petits pois*. (Literally 'little peas'.)

poisson (*Fr.*) (masc.) [pwah-so(n)] Fish.

poivrade, sauce (*Fr.*) [sohs pwahv-rwahd] Sauce consisting of a **mirepoix** mixed with **vinegar**, wine and crushed green **peppercorns**, traditionally served with game. (*See also* **chevreuil, salmis** *and* **venaison**.)

poivre (*Fr.*) (masc.) [pwahv-ruh] Pepper.

poivre, au (*Fr.*) [oh pwahv-ruh] With **pepper** or crushed **peppercorns**, e.g. **bifteck** *aux poivres*.

poivre vert (*Fr.*) [pwahv-ruh vair] Green **peppercorn**, e.g. **poulet** *aux poivres verts*.

polenta (*Ital.*) [puh-len-ta] Dough of thick porridge-like consistency made with cornmeal (maize flour). Cooked polenta can be fried or baked and served with meat dishes.

pollo (*Ital.* & *Span.*) Chicken.

pomme (*Fr.*) (fem.) [pom] Apple.

pomme de terre (*Fr.*) (fem.) [pom duh tair] Potato. (*De terre* means 'of the earth'.) Note that in compounds such as those below, the words '*de terre*' are omitted.

pommes Anna (*Fr.*) (fem.) [pom anna] Sliced potatoes cooked with butter in a pan to form a solid cake and served in slices. (Dish dedicated to Anna Deslions, socialite in the Second Empire, *c.* 1860.) (*Compare* **roesti**.)

pommes duchesse (*Fr.*) (fem.) [pom doo-shess] Potatoes **puréed** with egg yolk and butter, **piped** around the edge of a dish, or into small shapes, and baked in the oven.

pommes frites (*Fr.*) (fem.) [pom freet] Chipped potatoes deep-fried; French fries. (*See* **frite**.)

pommes noisettes (*Fr.*) (fem.) [pom nwa-zet] Small potato balls **sautéed** in butter. (*See* **noisette**.)

pommes sautées (*Fr.*) (fem.) [pom soh-tay] Cooked potatoes fried quickly, often with **shallots** or bacon to flavour. (*See* **sauté**.)

porcino (*Ital.*) (*pl.* porcini) [pore-cheen-oh(y)] **Mushroom** (*Boletus edulis*), esp. the dried mushroom. (*See also* **cep, morel, pine mushroom** *and* **truffle**.)

porridge (*Scot.*) [porij] Breakfast dish of oatmeal cooked with milk or water and served hot. (*Compare* **Bircher muesli**.)

port Kind of **fortified wine**, originally developed in Portugal, often served after dinner. (*See* Chapter 14.)

porterhouse Large beef steak cut from the **fillet** end of the **sirloin**. (*See also* **entrecôte, filet mignon, minute steak, rump** *and* **T-bone**.)

portion Amount of food served to guests in any particular menu item.

potage (*Fr.*) (masc.) [pot-ahzj] Soup, particularly thick (cream) soup. (*See* Chapter 2; *see also* **bisque, garbure, soupe** *and* **vichyssoise**.)

pot-au-feu (*Fr.*) (masc.) [pot oh fur] Classic dish of beef cooked slowly with vegetables in liquid to form a **broth**, which is eaten before the meat. (Literally 'pot for the fire', so 'stewpot' or 'casserole'.)

pot roast Joint of meat baked with vegetables and **stock** in a covered pan.

pot still Traditional **still** used to distil **malt whisky, cognacs** and **armagnacs**. The spirit is distilled in batches, unlike the 'continuous still' used to distil most modern **spirits**.

poulet (*Fr.*) (masc.) [poo-lay] Chicken, fowl.

poussin (*Fr.*) (masc.) [poo-sa(n)] Baby chicken.

praline [pray-leen] Confection of **caramelised** almonds or hazelnuts baked hard. It is added to some sweet dishes in the form of a fine, crushed powder. (Devised by Lassagne, chef to Marshal Duplessis-Praslin, *c.* 1630.)

Prashad Food eaten and blessed for Hindus.

preserve (1) (noun) Preserved fruit; jam, jelly.
(2) (verb) To prepare food for storage; to **pickle**.

preserved lemon Lemon **preserved** in salt and lemon juice. Flavourings, esp. **bay leaves**, cinnamon and **peppercorns**, can be added before sealing. After maturation, the lemon flesh is discarded and the soft skin is used as a flavouring.

pretzel Salted stick-like biscuit, shaped into knots and curls, served as an **appetiser**, with beer.

primi piatti (*Ital.*) (*pl.*) [preemy pee-atty] First courses; **entrées**, starters. (Literally 'first dishes'.) (*See also* **dolci** *and* **secondi piatti**.)

princesse, à la (*Fr.*) [ah lah pra(n)-sess] Garnished with asparagus tips.

Pritikin Style of cooking food named after Nathan Pritikin (1915–85), who maintained that by reducing **cholesterol** levels many diseases caused by a modern lifestyle could be prevented. The diet is low in fats and animal **protein**, has no added salt or sugar, and is high in fibre.

prix-fixe (*Fr.*) [pree fiks] Fixed-price meal; **table d'hôte** menu. (*Prix* means 'price'.)

profiterole (*Fr.*) (*fem.*) [prof-ee-tay-rol] Small **choux-pastry** bun with a sweet creamy filling served in chocolate sauce. Those filled with a cheese mixture or **savoury purée** are often used to garnish soup.

proof Old measure of alcohol in **liquor**, obsolete in Australia, NZ and UK, but still used in the USA (US 100° proof = 50% alc/vol). (*See also* **overproof**.)

proprietary The property of some person or body. A proprietary **liqueur–Grand Marnier**, for example–can only be made by the company that owns the right to make it. (*See also* **generic**.)

prosciutto (*Ital.*) [prosh-choot-oh] Raw **smoked** ham sliced very thinly. (*See also* **Parma ham**.)

protein Constituent of food needed for growth and repair, found particularly in meat and dairy products, but also in **soya beans**, **lentils** and peas.

provençale, à la (*Fr.*) [ah lah prov-ah(n)-saal] In the manner of Provence (province in S France); cooked with tomatoes and **garlic**.

prune (1) (*English*) Dried plum with black, glossy wrinkled skin. Prunes are stewed in liquor to make them edible or added to stews, esp. **tagines**. (2) (*French*) (*fem.*) Fresh plum, e.g. *tarte aux prunes* (plum tart).

pudding (1) Soft food item of ingredients mixed with or enclosed in flour or suet and usually steamed or baked, e.g. **bread and butter pudding**, **Christmas pudding** and **Yorkshire pudding**. (*See also* **black pudding** *and* **summer pudding**.)
(2) (*UK*) The **sweet** or **dessert** course of a meal. (*See* Chapter 2.)

puff pastry Delicate airy pastry rolled so that it forms thin layers when cooked. (*See* **pâte feuilletée**.)

pumpernickel (*Ger.*) [poom-purr-nickel] Solid, dark brown or black **rye** bread, often packaged for use as a base for **canapés**, cut into small circles or squares and sliced thinly.

punch Drink of wine or **spirits** with fruit juices or water flavoured with spices and often served hot. (*See also* **glühwein** *and* **mulled wine**.)

pungent With a strong taste or smell.

punt The hollow at the base of some wine bottles. It retains the deposit (if any) thrown by the wine and, esp. in the case of **champagne** bottles, may be used to hold the bottle while pouring. (*See* Chapter 15.)

purée [pew-ray] (1) (*Fr.*) (fem. noun) Any food **blended** to a smooth, thick paste, e.g. *purée de* **volaille**, 'purée of chicken'.
(2) (*Eng.*) (verb) To **blend** into a purée, e.g. puréed potatoes.

Puy lentils (*lentilles du Puy*) [lon-tea-yuh doo pwee] Tiny green **lentils** from Le Puy (France) that keep their shape when cooked.

quail Small **game** bird, nowadays farmed, usually served two birds to a guest.

quail egg Egg about one-third the size of standard hen's egg, often served hard-boiled as an **appetiser**. (*See also* Scotch egg.)

quandong/quondong (*Austral.*) [kwondong] Small bright-red fruit with a wrinkled stone containing an edible kernel; also known as native peach.

quenelle (*Fr.*) (fem.) [ken-nell] (1) (noun) Poached **dumpling**, resembling a small three-sided sausage, of **puréed** fish, egg white and **cream**. Veal or chicken is sometimes used instead of fish.
(2) (verb) To make a **quennelle**; to form food, e.g. mashed potato, pâté,

ice cream, into a **quennelle** shape using spoons.

queue de boeuf (*Fr.*) (fem.) [cu(r) duh be burf] Oxtail.

quiche (*Fr.*) (fem.) [keesh] Open pie or **flan**, filled with savoury **custard** holding other ingredients.

quiche Lorraine [keesh lor-aine] **Quiche** filled with savoury **custard** and bacon and topped with cheese. (Lorraine is a province in NE France.)

quince Hard, yellow, delicately scented, apple-like fruit that turns pink when cooked. Often used in pies and **puddings**; makes excellent **preserves**, esp. quince paste.

quinine Bitter alkaloid obtained from cinchona bark, once the principal treatment for malaria and used as flavouring, e.g. in Indian **tonic**.

race (1) Fenced access in a self-service **establishment**, which channels guests in a line past food and beverages towards the till. (Also called 'straight' race.)
(2) (**scrambled**) Self-service system where different types of food or beverage are available at different parts of the counter and are collected separately.

rack (of lamb) Joint of lamb cutlets cut from the rib **loin**. There are usually between four and six chops in the rack. (French **carré** *d'*agneau.)

raclette (*Fr.*) (fem.) [rack-let] (1) Curved scraping instrument for making butter-curls, etc. (*Racler* means 'to scrape'.)
(2) Cheese **fondue** made by melting a whole Swiss cheese and scraping off

pieces as they soften. It is usually eaten with plain potatoes, gherkins, **pickled** onions and black **pepper**.

radicchio (*Ital.*) [rad-eek-ee-oh] **Endive** with dark reddish-green leaves and distinctive white veins. Has pungent taste and mixes well with other salad greens. *See also* **mesclun.**

ragoût (*Fr.*) (masc.) [rag-oo] Stew or casserole, e.g. *ragoût d'*agneau.

rainbow chard *See* **silverbeet.**

raita (*Ind.*) [rye-eeta] Yoghurt blended with **garlic** and cucumber and served as a cooling **condiment** with spicy dishes.

ramekin [ram-uh-kin] Small baking dish for one serving; miniature **soufflé** dish. (*See* Chapter 3.)

rang (*Fr.*) (masc.) [rung] Rank, **brigade,** team (of dining-room service staff). (*See* **chef de rang.**)

rare Meat cooked for only a short time so that the blood runs; underdone. (*See also* **bleu, à point.**)

ras-el-hanout (*N Africa*) Spice mix, a blend of more than two dozen spices, esp. cardamom, coriander, cumin, ginger, nutmeg and turmeric. In Tunisia, ground rose petals are added to the mix. In Morocco, the blend is spicier. It is used to flavour **couscous** and meat dishes.

ratatouille (*Fr.*) (fem.) [rata-too-wee] **Provençal** vegetable stew containing tomatoes, onions, eggplant, **capsicums** and **zucchini,** flavoured with **garlic** and **pepper.**

ravière (*Fr.*) (fem.) [ravee-ayr] Oval or rectangular-shaped serving dish. (*See* Chapter 3.)

ravioli (*Ital.*) (*pl.*) [rav-ee-oh-lee] Squares of **pasta** filled with meat, **ricotta** cheese or a vegetable.

réchaud (*Fr.*) (masc.) [ray-show] Small stove used for table-side cooking or to keep food warm. (*Chaud* means 'hot'.) (*See* Chapters 3 *and* 12; *see also* **chafing dish, chaud** *and* **flambé.**)

réchauffé (*Fr.*) [ray-show-fay] (1) (masc. noun) Reheated dish; a dish made up of left-overs. (*See also* **émincés.**) (2) **réchauffé(e)** (adj.) Reheated. (*Chauffé* means 'heated'.)

red pepper (1) **Cayenne pepper.** (2) Variety of **sweet** (bell) **pepper** or **capsicum.**

reduce To thicken **stock** or liquids such as white wine or **vinegar** by boiling rapidly without a lid, thus reducing the quantity through evaporation.

refried beans Mashed beans, esp. pinto or black beans, which, once boiled, are fried with onions and spices. Refried beans are served with **tortillas, tacos** and similar Mexican cornbreads. (Spanish *frijoles refritos.*)

reggiano (*Ital.*) [redj-ee-ah-noh] *See* **grana** *and* **parmigiano.**

rehoboam [re-huh-boh-um] Large bottle of **sparkling** wine containing the equivalent of six standard (750mL) bottles. (Rehoboam was King of Judah in the 10th century BC.) (*See* **champagne bottle sizes.**)

relevé (*Fr.*) [ruh-lev-ay] (1) (masc. noun) Meat course in the **classic menu** following the **entrée,** usually a dish carved from a joint. (2) (adj.) Cooking term meaning 'highly seasoned'.

relish Condiment, **sauce** or **pickle**, usually eaten with plain food.

rémoulade (*Fr.*) (fem.) [ray-moo-laad] **Mayonnaise** sauce seasoned with gherkins, **capers**, **anchovies** and **herbs**.

rendang daging (*Indon.*) [rendan(g) dahging] Dry-fried beef **curry** cooked in **coconut milk** and spices.

reservation The booking of a table for a meal or bed for the night; the place thus booked.

restaurateur (*Fr.*) (masc.) [res-tara-tur] Manager or owner of a restaurant; the licensee. (Note: *not* restaura*n*teur.)

Rhine riesling Former Australian name for the popular white wine grape variety, now known simply as **riesling**. It is used to make **hock**. (*See also* **riesling**.)

rice paper Edible paper made from the pith of an oriental tree and used in Asian cooking. (*See also* **goi cuon**.)

rice vinegar Mild **vinegar** used in Asian cooking as a **pickling** agent, or to give an acidic or 'sour' taste to dishes. (*See also* **balsamic vinegar, malt vinegar, vinaigrette** *and* **wine vinegar**.)

ricotta (*Ital.*) White, moist, light cheese made mostly from the whey of milk. (*See also* **mascarpone**.)

riesling (*Ger.*) [reece-ling] Grape variety used to make white wines. (*See* Chapter 14. *See also* **auslese, hock, Rhine riesling** *and* **spätlese**.)

rillettes (*Fr.*) (fem. pl.) [ree-yettuh] **Minced meat**, usually pork, cooked in its own fat and made into a **pâté**. Served as cold **hors d'oeuvre**.

risoni (*Ital.*) [ri-zohn-y] Tiny balls of **pasta** rolled to resemble rice. (Literally 'little rice'.)

risotto (*Ital.*) [riz-ott-oh] Medium-grained rice, e.g. **arborio**, carnaroli or vialone, cooked in **stock** until the liquid is absorbed by the rice. Various vegetables, meat or fish might be added to make, e.g. mushroom risotto or shellfish risotto. (*See also* **risotto alla milanese** *below*.)

risotto alla milanese [riz-ott-oh ala mil-an-ayz-ay] Rice cooked in **stock** with onions, flavoured with **saffron** and **parmesan** cheese. (*Milanese* means 'in the style of Milan', city in N Italy.)

ristretto (*Ital.*) *See* Chapter 16.

rocket Pungent **cress** similar to spinach (*Eruca sativa*) used raw in **mesclun** mix for salad. Also known by its Italian name, **arugula**.

rock lobster Salt-water **crayfish** or southern rock **lobster**. (French langouste). (*See also* **bug**.)

roe Fish eggs. (*See also* **caviare** *and* **lumpfish roe**.)

roesti/rösti (*Swiss Ger.*) [rur-stee] Potato cake; grated potatoes formed into a cake and fried in butter. (*Compare* **pommes Anna**.)

rognon (*Fr.*) (masc.) [roh-nyo(n)] Kidney.

Romanov/Romanoff (*Fr.*) [roam-anoff]
(1) **Garnish** of cucumber stuffed with **duxelles**.
(2) Creamed-potato cases filled with mushrooms in a **velouté** sauce, seasoned with **horseradish**.
(3) Strawberries **macerated** in **curaçao** served with **crème chantilly**.

(Romanov was the name of the dynasty that ruled Russia from 1613 to 1917. Many French chefs dedicated dishes to the Imperial family.)

roquefort (*Fr.*) [rok-fore] Smooth and creamy **blue-veined** ewe's-milk cheese. (Roquefort is a district in S France.) (*See also* **dolcelatta, Gippsland blue, gorgonzola** *and* **stilton.**)

rosé (*Fr.*) [rose-ay] Pink wine made by removing the skins of black (red) grapes early in the **fermentation** process, resulting in a lighter colour and flavour than in an ordinary red wine.

rosella (*Austral.*) Fleshy red seed pod of *Hibiscus heterophyllus,* usually stewed and made into a **preserve.**

rosewater Sweet flavouring **essence** distilled from rose petals.

Rossini *See* **tournedos Rossini.**

rosso Sweet, pink style of **vermouth.**

Rosso Antico Proprietary Italian **apéritif** infused with aromatic herbs. (Literally 'Ancient Red'.)

rösti *See* **roesti.**

roti (*Ind.*) [roh-tee] General term for bread, particularly flat unleavened bread such as **chapati**; with **dahl,** it is the staple food of northern India.

rôti (*Fr.*) (masc.) [roh-tee] Roast meat, e.g. *rôti de* **boeuf.** *Rôtis* (pl.) is a course in the **classic menu.**

rotisserie Device for roasting meat on a rotating spit so that it browns evenly.

rouille (*Fr.*) (fem.) [roo-ee] Provençal sauce of **garlic,** oil and **stock** with red **chillies** and **saffron** served with fish dishes, esp. **bouillabaisse.** (Literally 'rust' from its colour.)

roulade (*Fr.*) (fem.) [roo-lahd] Stuffed roll of food. A savoury roulade may be a thin piece of meat stuffed and rolled; a sweet one a chocolate sponge filled with **cream** before rolling.

roux (*Fr.*) (masc.) [roo] Blend of fat and flour used to thicken sauces. (*See* **béchamel.**)

RSA Responsible service of alcohol. (*See* Chapter 15.)

RTD Ready-to-Drink; mixed drinks available ready-mixed in bottles or cans.

rum Spirit distilled from sugar-cane or molasses. Rum comes in various styles and colours, ranging from colourless (such as **Bacardi**) to dark brown Jamaica rum.

rum baba *See* **baba.**

rump steak Tender cut of beef taken from behind the **loin.** (*See also* **entrecôte, filet mignon, porterhouse, sirloin** *and* **T-bone.**)

russe, à la (*Fr.*) [ah lah rooss] In the Russian style; often shellfish in an **aspic** jelly with **mayonnaise,** or a **chaudfroid** sauce served with a **Russian salad.** (*See also* **charlotte russe.**)

Russian salad Salad of **diced** cooked vegetables mixed in **mayonnaise** and flavoured with either caraway or **pickles.**

Russian tea Tea served without milk in a silver holder with a lemon slice.

rye (1) **Cereal** used for making bread; rye bread is rather dark in colour and slightly bitter in taste.
(2) Rye **whiskey;** American whiskey distilled (mainly) from fermented rye.

sabayon (*Fr.*) (masc.) [sab-eye-o(n)]
Mixture of egg yolks, white wine and
sugar whipped together to form a rich
foamy dessert, usually served just
warm. Similar to Italian **zabaglione**.

sabayon sauce Savoury sauce made from
egg yolks blended with alcohol
(particularly **sparkling wine**) and
usually served with fish.

sablé (*Fr.*) (masc.) [sab-lay] Kind of
shortbread. (Literally 'covered with
sand'.)

saddle Large cut of meat from the top of
the back of the animal. (*See also* **loin**.)

saffron Dried stigma of a kind of crocus
flower used as a spice for flavouring.
It colours food bright yellow, and is
also used as a dye.

saignant(e) (*Fr.*) [say-nyo(n) (tuh)] **Rare**
and rather bloody (steak, etc.) but not
'blue'. (*See also* **bien cuit**, **bleu** *and*
point, à.) (Literally 'bleeding'.)

sake (*Jap.*) [sah-kay] Fermented,
colourless rice wine, usually served
warm. (*See also* **mirin**.)

salad dressing *See* **French dressing,
Italian dressing, mayonnaise,
Thousand Island dressing** *and*
vinaigrette.

salade (*Fr.*) (fem.) [sall-ahd] Salad. (Italian
insalata; Spanish **ensalada**.) (*See also*
**Caesar, classic menu, fattoush,
mesclun, niçoise, panzanella,
Russian, tabbouleh** *and* **Waldorf**.)

salamander Grill that heats from above,
used for browning, etc.

salami (*Ital.*) [salaamee] Highly seasoned
sausage eaten cold. (*See also* **kabana,
King Island, mortadella, pepperoni,
charcuterie** *and* **smallgoods**.)

salmanazar Large bottle size for
sparkling wine, equivalent of 12
standard (750mL) bottles. (Salmanazar
was the Assyrian king who conquered
Israel in 721 BC.) (*See also*
champagne bottle sizes.)

salmis (*Fr.*) (masc.) [salmee] **Ragoût of
game** bird or duck, often finished at
table. (*See also* **salmis, sauce** *below*.)

salmis, sauce (*Fr.*) [sohs salmee] Sauce
served with a **salmis** made with the
cooking juices diluted with wine
mixed with an **espagnole** sauce.

salmonella Group of **bacteria** which can
cause food poisoning.

salpicon (*Fr.*) (masc.) [sal-pee-ko(n)]
Diced pieces of meat, etc. held in a
sauce and used to fill **canapés, vols-
au-vent**, etc.

salsa (*Ital., Span. & Mex.*) [sals-a] (1)
Sauce. (*See below*.)
(2) Combination of chopped and
usually raw vegetables, served as a
fresh **chutney**, e.g. **salsa verde**.

salsa di pomodoro (*Ital.*) [sals-a dee
pom-uh-dorr-oh] Tomato sauce.
(Literally 'sauce of tomato'.)

salsa verde (*Ital.*) [sals-a vairduh] Thick
green sauce of **anchovies, parsley,
capers**, olive oil and lemon juice or
vinegar blended with bread.
Cucumber, basil or onion can be
added. (Literally 'green sauce'.)

salsa verde cruda (*Mex.*) [salsa vairduh
crooda] Uncooked green-tomato sauce
served with **tacos, nachos, tortillas**
and **enchiladas**. (Literally 'raw' or
'unripe' 'green sauce'.)

saltimbocca (*Ital.*) [saul-teem-boh-kah]
Thin slices of **veal** and **smoked ham**

rolled with sage leaves, **sautéed** and simmered in wine. (Literally 'jump into mouth'.)

salver Tray, usually round and made of silver, on which drinks, etc. are presented.

sambal/sambol (*Indon.*) [sam-bahl] **Condiment** usually containing **chillies** and onion, served with curries. (*See also* **sambal ulek** *below.*)

sambal ulek/oelek (*Indon.*) [sam-bahl uluk] Hot **chilli** paste used as a seasoning or **condiment**.

samosa (*Ind.*) [samohsa] Small pastry with spicy filling deep-fried and served as a **savoury**.

sashimi (*Jap.*) [sash-eemee] Raw fish arranged in thin bite-sized pieces and served with **daikon radish, tamari** and **wasabi**. (Literally 'pierced flesh'.)

satay (*Malay.*) [sah-tay] Cubes of meat or fish on bamboo skewers, grilled on a charcoal grill, and served with **satay sauce**. (*See* **kebab**.)

satay sauce (*Malay.*) Thick sauce made from crushed peanuts and lemon juice, often served with **satay**.

sauce *See* **anchoiade, béarnaise, béchamel, beurre blanc, beurre noisette, beurre meunière, black bean, bolognaise, capricciosa, carbonara, charcutière (à la), chasseur, crème anglaise, Cumberland, custard, demi-glace, espagnole, florentine, French dressing, ganache, hoi sin, hollandaise, horseradish, ketchup, Kilpatrick, lyonnaise, maître d'hôtel, mayonnaise, meunière, mint, mole, mornay, mousseline, nam pla, nam prik, newburg, normande, nuoc** cham, **nuoc mam, poivrade, relish, rémoulade, rouille, sabayon, salmis, salsa di pomodoro, salsa verde, satay, shoyu, soubise, soya, sugo, suprême, sweet-and-sour, Tabasco, tamari, tartare, teriyaki, thermidor, Thousand Island, velouté, venaison, vinaigrette, Worcestershire** *and* **zingara**. (*See also below and* **crème (à la), mask, nappé, roux**.) (French *sauce* (fem.) [sohs] as in *sauce béarnaise;* Italian and Spanish **salsa**.)

saucière (*Fr.*) (fem.) [soh-see-ayr] Sauce or gravy boat. (*See* Chapters 3 and 8.)

sauerkraut (*Ger.*) [sour krout] **Pickled** white cabbage. (Literally 'sour cabbage'.)

saumon (*Fr.*) (masc.) [sow-mo(n)] Salmon.

sauté (*Fr.*) [soh-tay] (1) (masc. noun) Dish of quickly fried food.
(2) (adj.) (*sauté(e)*) Lightly fried, e.g. **pommes** *sautées*.
(3) (*Eng.*) (verb) Lightly fry—a 'sautéed' mixture, 'sautéed in butter', etc.

sauternes [so-tairn] Lusciously rich, sweet white **dessert wine** from Sauternes, district near Bordeaux, SW France. (*See* Chapter 14.)

sauvignon blanc [so-veenee-o (n) blo(n)] Grape variety used to make dry white wines, including **fumé** blanc.

savarin (*Fr.*) (masc.) [sav-a-ra(n)] Large ring-shaped cake made with yeast dough served soaked in **rum** and filled with **cream** or **custard**. (Named after Jean-Anthelme Brillat-Savarin, 1755–1826, gastronome and writer.)

savoury (1) (adj.) Salty, piquant, tasty.
(2) (noun) Small savoury item; **canapé**. (*See also* **classic menu**.)

scald (1) To dip fruit or vegetables in boiling water to remove impurities or prepare them for skinning or freezing. (2) To heat (particularly milk) until just at the point of boiling.

scallion Green (immature) onion; spring onion.

scallop (skol-op] (1) (noun) **Mollusc** or shellfish (genus *Pecten*) contained in a big fan-like shell on which it is often served. The 'meat', which is opaque and browny-white when raw, with a 'coral' attached, turns white when cooked. (*French* **coquille**.)
(2) (verb) To decorate with semi-circular shapes; to **score** with such a pattern.

scalloped potatoes (*USA*) [skol-lopt] Sliced potatoes baked with milk. (*See* **pommes Anna**.)

scallopina (*Ital*.) (*pl*. **scallopini**) [ska-lop-een-ee] **Escalope** or thin slice of meat lightly crumbed and **sautéed**; traditionally, a sauce of **marsala**, tomato and **cream** is served with it.

scampi Variety of large prawn. (Plural of the Italian *scampo;* in French, **langoustine**.)

schnapps (*Ger*.) [shnaps] N European flavoured distilled **spirit** made from grain or potatoes, similar to **hollands gin**.

schnitzel (*Ger*.) [shnit-sell] Thin slice of meat, usually chicken or **veal**. (*See* **Wiener schnitzel**.) (*Schnitzeln* means 'to cut', so schnitzel means 'a slice'.)

scone [skon] Bread-like small cake or bun served with jam and butter or thick **cream**. (*See also* **Devonshire tea**.)

score To cut lines or marks into a surface, esp. in skin or rind.

Scotch Scotch **whisky**.

Scotch egg (*UK*) Whole hard-boiled egg encased in sausage meat, which is crumbed and fried. It is served cold.

seafood Collective term for edible fish, shellfish in particular. (*See* **fruits de mer, marinara, pesce** *and* **poisson**.)

seal To sear the surface of meat by intense heat to seal in the juices.

sec (*Fr*.) (fem. **sèche**) [sek/sesh] Dry (when referring to wine). But note that the word *sec* is sometimes confusingly applied to drinks that are not really dry, e.g. **triple sec** and **champagne**. (*See* Chapter 14.)

secondi piatti (*Ital*.) [sekondy pee-atty] Second or main courses. (Literally 'second dishes'.) (*See also* **dolci** *and* **primi piatti**.)

sekt (*Ger*.) [sekt] German **sparkling wine**.

semi-dried tomato *See* **tomato, dried**.

semifreddo (*Ital*.) Ice cream, usually made with eggs and cream. It is smooth and creamy in texture and doesn't freeze rock hard. (*See also* **gelato, sorbet** *and* **water ice**.)

sémillon [say-me-yo(n)] Grape variety used to make crisp dry white wines.

semolina (*Ital*.) Coarsely ground grains of hard wheat or rice, used to make many **pasta** products and **couscous**.

service charge Charge added to a guest's bill to cover staff costs, in theory later distributed to the waiters.

service cloth A (white) cloth or **table napkin** used by the waiter in food service. (*See* Chapter 7.)

service gear Utensils used by the waiter in food service, esp. **silver service**, usually a tablespoon and fork. (*See* Chapters 3 *and* 8.)

service plate Large plate covered with a **service cloth**, used, for example, to carry cutlery to tables once customers are seated. (*See* Chapter 6.)

serviette *See* **table napkin**.

sesame seed Small seed from sesame plant, oily and nourishing, widely used in cooking, e.g. to make **halva** or sprinkled on loaves of bread. (*See also* **hummus** *and* **tahini**.)

set menu Menu allowing no choice to the guests, all items having been pre-selected by the **host**, typically used at **functions**. (*See* Chapter 2.)

sformato (*Ital.*) [s-form-ahtoe] Kind of **soufflé**, either sweet or savoury, served with accompanying sauce of meat or flavoured **cream**.

shake and strain Cocktail-mixing term. The ingredients are shaken together with ice in a cocktail shaker and strained through a **Hawthorn strainer** into the glass. (*See* Chapter 14.)

shallot Small brown or reddish-brown onion (*Allium ascalonicum*). The word is sometimes incorrectly used to mean **spring onion**. (*Also known as* **eschalot**; French *èchalote*.)

shank Cut of meat, esp. veal or lamb, from the leg. (*See also* **carré**.)

shaslik Kebab.

sherbet (1) Sweet, flavoured effervescent drink.
(2) (*USA*) Water ice or **sorbet**.

sherry **Fortified wine** made from white grapes, originally developed in Jerez in Spain. (*See* Chapter 14.)

shiraz [shi-raaz] Grape variety, formerly also called hermitage, suiting warm conditions. It produces robust red wines that benefit from long maturing.

shirred egg Baked egg.

shish kebab (*Turk.*) [shish kuh-baab] Small pieces of meat, usually lamb, cooked on a skewer. (*See* **kebab**.)

shitake (*Jap.*) [shit-ah-ky] Mushroom (*Lentinus edodes*) with a strong flavour. The caps are brown and the stems tough but flavoursome. (*See also* **cep**, **morel**, **mushroom**, **oyster mushroom**, **pine mushroom** *and* **truffle**.)

shooter Layered **cocktail** without ice served in a small glass.

shortbread (*UK*) Crumbly sweet biscuit of flour (originally oatmeal) and butter.

shortcake Dessert of sweet pastry or cake sandwiched with fruit, esp. strawberries, and **cream**.

shortcrust pastry Basic pastry mix of flour, fat and water. It may be sweetened with sugar, and egg or lemon juice may replace the water. (*See* **pâte brisée**, **pâte sablée** *and* **pâte sucrée**.)

shoyu (*Jap.*) [show-yu] Popular style of Japanese **soya sauce**; it is less salty and lighter than the most common Chinese varieties.

shuck (*USA*) To remove from shell or cob.

shwarma/chawarma (*Leb.*) [sha-warma] *See* **döner kebab**.

sideboard Table or chest for holding plates and glass, and from which food is served; a waiter's **station**. (*See* Chapters 1 *and* 4.)

silverbeet/Swiss chard Glossy, green-leaf vegetable with white stalks that look like celery. The stem and the leaf are edible but are usually prepared separately. The leaves are often used as a substitute for spinach. Rainbow chard has brightly coloured stems of red, yellow and orange. (French *blettes*.)

silver service Formal style of food service whereby food items are transferred by the waiter to the guests' plates from service dishes at the table. (*See* Chapter 8.)

silverside Boneless hindquarter of beef, usually salted and **pot roasted**.

sirloin Choice joint of beef taken from the upper **loin**, which includes the **fillet**. It is often boned and rolled with the fillet removed. It can be cut into steaks. (*See also* **entrecôte**, **filet mignon**, **porterhouse**, **rump** *and* **T-bone**.)

skillet Frying pan; cast-iron pot.

skordalia (*Grk*) Potato **puréed** with garlic and served as a dip, esp. on a **mezze** plate, or as an accompaniment to meat dishes.

slivovitz Plum **brandy** from the Balkans. (*Sljiva* means 'plum' in the Serbian and Croatian languages.)

sloe gin Liqueur of sloe berries soaked in gin.

smallgoods Cooked meats and meat products, sausages, salamis, **pâtés**, etc. (*See also* **charcuterie**.)

smoke To **preserve** or **cure** food, e.g. ham, salmon or turkey, by drying it above smoking wood or peat with mixtures of sugar, spices and **liquors** added to impart special flavours.

smoothie Beverage of blended fruit mixed with milk and honey. (*Compare* **lassi**.)

smorgasbord (*Swed.*) Self-service buffet (*see* Chapter 10). (Literally 'lump of butter' (*smörgås*) + 'table' (*bord*). In Scandinavia, a *smörgåsbord* is a very substantial meal of several courses, esp. including dishes of herrings.)

smørrebrød (*Dan.*) [shmuruh-brur(t)] Open sandwich.

snow pea *See* **mange-tout**.

snow pea shoots Small pale-green shoots from the **snow pea** plant, usually served mixed in green salads.

soba (*Jap.*) Pale brown, thin **noodles** with a distinct taste of **buckwheat**, from which they are made. (Soba means 'buckwheat'.) (*Compare* **udon**.)

soda (**water**) Water aerated with carbon dioxide.

soffritto (*Ital.*) Mixed diced vegetables and herbs (onion, garlic, carrots, etc.) browned in fat and used as foundation for soups and stews. (*Soffriggere* means 'to fry lightly'.) (*Compare* **mirepoix**.)

sommelier (*Fr.*) (masc.) [som-ell-ee-ay] Expert wine waiter.

sorbet (*Fr.*) (masc.) [sor-bay] Light soft frozen mixture of **puréed** fruit or fruit juice with some **liquor** and **sugar syrup**; **water ice**. Sorbets are often served between courses to freshen the **palate**. (*See also* **classic menu**.)

sorrel [sorrul] Green-leaf vegetable with a slightly bitter taste. It is used to make soup and, when young, mixed in green salads. (French *oseille*.)

soubise (*Fr.*) (fem.) [soo-beez] An onion sauce or **purée** of onion and rice.

soufflé (*Fr.*) (masc.) [soo-flay] Sweet or savoury sauce mixture to which beaten egg whites have been added, which cause it to rise once baked. It is almost always served hot, but a cold sweet soufflé is made from fruit **purée** to which egg white is added, e.g. *soufflé aux* **fruits** (fruit soufflé), *soufflé au* **fromage** (cheese soufflé), *soufflé au chocolat* (chocolate soufflé).

soupe (*Fr.*) (fem) [soop] Soup, esp. those containing bread. (*See* Chapter 2. *See also* **bisque, consommé, garbure, julienne, marmite petite, pistou, potage** *and* **vichyssoise.**)

sour cream *See* **cream, soured.**

sourdough Piece of dough left over from a previous batch and used as a 'starter' to leaven a new batch. Sourdough bread has a crisp crust and soft inner crumb, a slightly sour taste, and is usually baked by traditional methods, possibly in a wood-fired oven.

sous chef (*Fr.*) [soo shef] Deputy chef. (*Sous* means 'under'.)

Southern Comfort US **proprietary liqueur,** based on **bourbon whiskey** with peach flavourings.

souvlaki (*Grk*) [soov-lahkee] Grilled or barbecued meat, esp. lamb, wrapped in **pita** bread with **humus** and salad; meat cooked on a skewer. (*See also* **kebab.**)

soya/soy bean Round bean, the same size as a pea, rich in protein, used to make **tofu,** etc.

soya/soy sauce Vital ingredient and popular **condiment** of E Asian cuisine, made from **soya beans** and water with other ingredients such as wheat, **vinegar,** salt, etc. There are many different kinds of soya sauce, including the Japanese **shoyu** and **teriyaki sauce.**

spaetzle/späzle (*Ger.* & *Alsace*) [spaytz-luh] Small poached **dumplings** of flour, eggs and cream, normally used to **garnish** meat dishes or served as an **entrée,** *au gratin.*

spaghetti (*Ital.*) (*pl.*) [spag-ay-tee] **Pasta** cut into thin strings. (Literally 'little strings'.) (*See also* **carbonara.**)

spanakopita (*Grk*) [spana-kohpeeta] Pasty made of **filo pastry** containing spinach and cheese.

Spanish olive Particularly mild-tasting green **olive** suitable for use in **cocktails.**

spare ribs The rib bones of pork or beef, usually **marinated** and baked.

sparkling wine Wines made effervescent (bubbly) by the natural action of the carbon dioxide gas produced by **fermentation. Champagne** is the most famous kind of sparkling wine. (*See* Chapter 14. *Compare* **carbonated wine.**)

spatchcock Dressed young chicken or **game** bird that is split open and grilled or fried without being allowed to hang and mature.

spätlese (*Ger.*) [shpayt-layz-uh] Sweet dessert wine made from late-picked **riesling** grapes. (*See also* **auslese** *and* **riesling.**) (*Spätlese* means 'late-picked'.)

späzle *See* **spaetzle.**

spelt flour Flour made with an ancient grain that makes a soft, nutty flour.

Spelt flour contains less **gluten** than standard flour. Also known as **farro**.

spirits General term for **distilled** alcoholic liquors, including **brandy**, **gin**, **rum**, **tequila**, **whisky** and **vodka**. (*See* Chapter 14.)

split Small or half-bottle (285mL) of an effervescent drink such as **soda** or **tonic**.

split pea Dried pea; split, skinned, dried pulse used esp. in pea and ham soup. (*See also* **lentil**.)

spoom Sorbet stirred with **meringue** so that it foams. (In Italian, *spuma* [spoo-mah].)

spring roll Deep-fried Chinese-style **pancake** parcel containing minced meat and vegetables. (*See also* **cha gio** *and* **goi cuon**.)

spumante (*Ital.*) [spoo-man-tay] Sparkling; Italian **sparkling wine**.

squab [skwob] Young pigeon.

staphylococcus **Bacterium** causing food poisoning; found in human throats and noses, and septic cuts.

starch **Carbohydrate** found in cereals and potatoes.

station (1) Section of the dining-room for which a waiter or team of service staff is responsible. (*See* Chapter 1.) (2) Table or **sideboard** where a waiter keeps the equipment necessary for service; a waiter's work station. (*See* Chapter 4.)

steak tartare *See* **tartare** *and* **tartare steak**.

steamboat Style of Chinese cooking where a pan of **broth** is placed on a heat source in the centre of the table into which diners dip raw food until it is cooked. At the end of the meal this **broth**, which has become quite rich, is eaten as soup.

Stelvin screwcap Screw top on a wine bottle. (*See* Chapter 15.)

still Apparatus used to distil **spirits**. (*See* **distillation**, **pot still** *and* Chapter 14.)

still area/stillroom Store or **pantry** for small items of food and beverage needed for service but not provided by the kitchen or larder.

stilton (*UK*) Rich, creamy **blue-veined** cheese, now usually served on cheeseboards. Traditionally a whole stilton was presented and scooped out with a spoon, with **port** filling the hollow so created. (Stilton is a village in central England.) (*See also* **dolcelatta, Gippsland blue, gorgonzola** *and* **roquefort**.)

stir and strain Cocktail-mixing term. Ice is placed in a mixing glass, the **liquors** are added and stirred until cold, then strained into the serving glass using a **Hawthorn strainer**.

stir fry (1) (verb) E Asian method of preparing food whereby the ingredients are chopped into small pieces and cooked over a fast heat in a **wok** while being stirred. (2) (noun) A stir-fried dish.

stock **Broth** made by simmering bones and vegetables; used as the basis for **sauces**, gravies, soups and stews.

stout Strong dark **beer** brewed with roasted **malt**.

stracci (*Ital.*) [strah-chee] **Pasta** cut into wide flat ribbons (literally 'rags').

Stroganov *See* **boeuf Stroganov**.

strudel [stroo-del] Dessert with case of **filo pastry**, e.g. apple strudel (German *Apfel Strüdel. Strüdel* means 'whirlpool'.)

sucré(e) (*Fr.*) (adj.) [sookray] Sugared. (*Sucre* means 'sugar'.)

suet [sooet] Solid fat from around the kidney of beef or mutton, fresh or dried, used in traditional English **Christmas** (plum) **pudding** and in steamed pastry dishes, e.g. steak and kidney **pudding**.

sugarbark (*Austral.*) Garnish of mixed sugars baked until crisp and golden. (Recipe created by Jean-Paul Bruneteau of Rowntrees The Australian Restaurant.)

sugar syrup Sweet **liquor** made of sugar and water, boiled and used when making desserts or poaching fruit and in some **cocktails; gomme syrup**.

sugo (*Ital.*) [soo-goh] **Sauce**, esp. tomato; gravy; juice.

sukiyaki (*Jap.*) [sooky-yahkee] Small lumps of meat cooked in hot oil at table with diners helping themselves from the common pot, **fondue**-style. (*Suki* means 'scoop' and *yaki*, 'grilled'.)

sumac (*Mid East*) Dark red spice with a sour, lemony taste, used to flavour savoury dishes, esp. **kebabs**.

summer pudding (*UK*) Dessert of white bread soaked with sweetened berries, redcurrants and blackcurrants and turned upside down to serve.

sundae [sunday] Ice cream dessert with topping of fruit, nuts or syrup.

sun-dried tomato *See* **tomato**, dried.

supper (1) Insubstantial late evening snack or very light meal taken after dinner or tea.
(2) (*UK*) Informal evening meal replacing a more formal dinner.

suprême (*Fr.*) (masc.) [soo-praym] (1) Breast and wing of poultry or **game**, e.g. *suprêmes de volaille* (chicken breasts in sauce).
(2) **Fillet** of large fish. (*See also* **suprême, sauce**.)

suprême, sauce (*Fr.*) [sohs soo-praym] Rich sauce containing **cream** and butter.

sushi (*Jap.*) [soo-shee] Bite-sized rice balls or sandwiches flavoured with **vinegar** and topped with **marinaded** raw fish or some other delicacy, often wrapped in **nori**.

Suzette *See* **crêpes Suzette**.

sweet (1) (adj.) Tasting of sugar; opposite of sour.
(2) (noun) Small item of confectionery, usually mainly of sugar or chocolate.
(3) (noun) The sweet or **dessert course** of a meal. (*See* Chapter 2; *see also* **entremets** *and* **pudding**.)

sweet-and-sour In Chinese cuisine, dishes cooked with a sauce made from sugar, **soy sauce** and **vinegar** or lemon, usually thickened with cornflour.

sweetbreads Pancreas or thymus of an animal (usually **veal** or lamb) when used in cooking. (*See* **abats**.)

sweet corn/maize *See* **corn**.

sweetmeat Sugar-coated confection; small fancy cake. (*See also* **fondant** *and* **petit four**.)

sweet pepper Capsicum or bell pepper; green, red, yellow or purple pepper.

Vegetable (*Capsicum annuum*) related to the hot **chilli**, but not the pepper spice. (*See* **pepper**.)

Swiss chard *See* **silverbeet**.

syllabub (*UK*) [silla-bub] **Dessert** of sweetened **cream** with lemon juice and wine or **brandy**.

Szechwan [sitch-wahn] One of the major styles of Chinese cuisine, and typically hot and spicy. (*See also* **Cantonese, Fukien, Honan** *and* **Peking**.) (Szechwan, now spelt Sichuan, is a province in NW China.)

Tabasco [tab-ass-coh] Spicy **proprietary** sauce made from the pungent tabasco **chilli** pepper. Sold in small bottles, it is used to season meat and egg dishes and **cocktails**. (Tabasco is a state in Mexico.)

tabbouleh/tabouli (*Arab.*) [taboo-lee] Salad of **cracked wheat**, onion, **mint**, tomato *and* parsley.

table d'hôte (*Fr.*) [tahbl doht] Type of menu offering limited choice at a fixed price for the whole meal. (*See* **prix fixe** *and* Chapter 2.)

table linen Tablecloth and **table napkins**; napery (not necessarily made from linen cloth).

table napkin Serviette; a cloth placed on diners' laps and used to protect their clothes and to wipe their fingers and mouths. (*See also* **service cloth**.)

table wine Wine that is drunk with a meal; unfortified still wine. (*See* **fortified wine, sparkling wine** *and* **wine**; *see also* Chapter 14.)

taco (*Mex.*) [tahkoh] Sandwich in a crisp **corn pancake** or taco shell; small **tortilla** wrapped around a filling, usually of meat and/or beans and cheese. (*See also* **refried beans**.)

tagine (*Moroccan*) [ta-zjeen] **Casserole** of meat and/or vegetables with fruit and spices.

tagliatelle (*Ital.*) (*pl.*) [tie-ya-telluh] Long ribbons of **pasta**, very like fettuccine. (*Tagliare* means 'to cut'.)

tahini/tahina (*Arab.*) [tuheen-ee] **Sesame seed** paste, an ingredient of **hummus** and **baba ghanoush**. (*Tahina* means 'ground' or 'crushed'.)

tamari (*Jap.*) [tam-ahr-y] Thick, reddish-brown **soy** sauce often served with **sashimi**.

tamarind (*Arab.*) Fruit of the tamarind tree (*Tamarindus indica*) the black pulp of which is used in Indian cuisine. It is rarely available fresh in Australia, but is compressed in a block or bottled as a paste.

tandoor (*Ind.*) (noun) [tun-door] Open-topped clay oven. *See also* **tandoori** *below*.

tandoori (*Ind.*) (adj.) [tun-door-ee] Food that has been cooked in a **tandoor**, e.g. tandoori chicken.

tannin Ingredient of wine, esp. full-bodied red wine, derived from pips and stalks which gives a drying, furry sensation to tongue and gums.

tapas (*Span.*) [tap-ass] Substantial cocktail snacks or **appetisers**. (*See also* **hors d'oeuvres**.)

tapas bar Brasserie or wine bar at which **tapas** are served.

tapénade (*Fr.*) [tapay-nahd] Provençal paste of **olives** blended with **anchovies, capers, garlic** and oil.

tapioca [tap-ee-oh-ka] Rice-like grains (or 'pearls') of dried cassava baked with milk and served as a pudding. (*See also* **bubble tea.**)

taramasalata (*Grk*) [tarra-mah-sall-ah-tah] Pale pink dip made of fish **roe** blended with olive oil and **garlic**, usually served as part of the **hors d'oeuvres** or **meze.**

taro Root vegetable (*Colocasia antiquorum*) rather like a hairy potato. Its taste resembles a dry sweet potato. (Also known as dasheen.)

tarragon Plant with long narrow green leaves used as a flavouring herb. French tarragon is much finer than the Russian variety. Tarragon is one of the ingredients of **fines herbes.**

tartare, à la (*Fr.*) [ah lah tar-tar] Literally 'in the manner of the Tartars', the once-terrifying nomadic hordes of Central Asia. (*See below.*)

tartare sauce [tar-tahr] **Mayonnaise** sauce made from hard-boiled egg yolks, onion, **chives** and **capers**; served with fish.

tartare steak/steak tartare [tar-tahr] Minced raw beef garnished with **capers**, minced onion and parsley and served with a raw egg yolk. (In French, **bifteck** *à la* **tartare.**)

tarte (*Fr.*) (fem.) [tart] Tart or open pie, e.g. **tarte tatin.**

tarte tatin (*Fr.*) [tart tata(n)] Upside-down apple tart. (The Tatin sisters, restaurateurs, popularised the dish in the early 1900s.)

tartuffo (*Ital.*) Espresso coffee served in a full-sized coffee cup poured over ice cream. (*Compare* **affogato.**)

T-bone Beef steak on a T-shaped bone cut from the **fillet** end of the **sirloin.** (*See also* **entrecôte, filet mignon, porterhouse, rump** *and* **sirloin.**)

tea (1) Mildly stimulating beverage made by infusing the dried leaves of the tea bush in boiling water. The two main styles are green (or **China**) tea and black (or Indian) tea. There are many varieties, e.g. Assam, **Darjeeling, Earl Grey, English Breakfast, jasmine**, etc. Some drinks are called 'tea' although they are not made with tea leaves, e.g. **herbal teas.** (*See also* **ayurvedic tea, bubble tea, chai, iced tea** *and* **tisane**, *and* Chapters 14 *and* 15.) (2) Meal at which tea is served, e.g. afternoon tea; the main (informal) evening meal.

tempeh (*Indon.*) [tem-pay] Fermented **soya beans** formed into a block. The white mould crust is edible. Tempeh absorbs other flavours and is usually **marinated** before being fried or grilled. (*Compare* **tofu.**)

tempura (*Jap.*) [tem-poor-ah] Strips of fish, shelled prawns or vegetables fried in a light **batter** and served very hot, usually with grated **daikon radish.**

teppanyaki (*Jap.*) [tep-an-yak-ee] Meat, fish and vegetables fried at the table. (Literally 'table-grilled'.)

tequila Mexican **spirit**, of which there are two varieties, white and gold, distilled from the cactus-like **mezcal azul** plant. Tequila is an ingredient of many **cocktails**, e.g. **Margarita.** (*See also* **mezcal.**)

teriyaki (*Jap.*) [terri-yah-kee] Meat or fish **marinated** in **teriyaki sauce** and grilled. (Literally 'sunshine-grilled'– glazed and grilled.)

teriyaki sauce Soya sauce brewed with sweet rice wine, **vinegar**, sugar and spices.

terrine (*Fr.*) (fem.) [teh-reen] (1) Deep oblong cooking dish with a tight-fitting lid. (Not to be confused with a **tureen**.)
(2) Chopped and minced meat of different varieties (or occasionally fish) mixed with vegetables and seasoning and cooked in a **terrine** until they form a 'loaf' which is sliced and served cold.

therapeutic diet A diet that can meet the nutritional needs of the sufferer of a medical illness or condition.

thermidor Recipe for **lobster** or **crayfish** served baked in their shells with a creamy sauce and hot **mustard**, often **au gratin**.

Thousand Island dressing (*Canada & USA*) Sauce of **mayonnaise** flavoured with **tomato sauce**, onion and **chilli**. (The Thousand Islands are in the north-east of Lake Ontario.)

thyme [time] Shrub with small pungent leaves used as a seasoning herb, esp. in soups, stews and stuffings. A component of the traditional **bouquet garni**, thyme is also a vital element in **Cajun** cooking.

Tia Maria Rum-based, coffee-flavoured **liqueur** from Jamaica.

tikka (*Ind.*) Pungent, spicy meat dish. Literally 'small piece of meat'.

timbale (*Fr.*) (fem.) [tam-bahl] (1) Cup-shaped mould.
(2) Food pressed into a timbale, served turned out upside-down.
(3) Piecrust baked in a timbale and filled with a creamy savoury mixture

(croûte *à timbale*).
(4) **Dessert** of fruit or a **cream** presented in a cup-shaped basket made from biscuit or pastry dough. *Une timbale* is 'a kettledrum'.

tira-mi-su (*Ital.*) [teera-mee-soo] Cake of **macerated** biscuits (**ladyfingers**) sandwiched with a **cream** made of sweetened **mascarpone** and separated eggs. (Literally 'pick-me-up'.)

tisane [tizann] *See* **herbal tea**.

tofu (*Jap.*) [toe-foo] Highly nutritious **curd** made from **soya beans**; it resembles a lump of soft cheese or firm **custard**. It is a versatile ingredient that tastes of little but absorbs the flavour of accompanying food or sauce. (*Compare* **tempeh**.)

tokay (1) Classic sweet white unfortified **dessert wine** of Hungary.
(2) Australian **fortified** dessert wine similar to **muscat** made from a variety of grape called tokay in Australia (but muscadelle in Europe).

tomato Versatile vegetable used in many cooked dishes and **chutneys**, and eaten raw in salads. Roma, also known as Italian plum, tomatoes are used fresh and for canning. Other tomato products include tomato juice, tomato **purée**, tomato paste and tomato **ketchup** (or sauce). Cherry tomatoes are a miniature variety. (*See also* **salsa di pomodoro**, **sugo** *and* **tomato, dried**. *Compare* **bush tomato**.)

tomato, dried Dehydrated **tomato**, either sun-dried, semi-dried or oven-dried. Sun-dried tomatoes are dark red-brown and shrivelled; semi-dried tomatoes still contain moisture and are bright red. They are preserved in oil. Oven-dried tomatoes are cooked slowly in the oven.

Tom Collins Cocktail containing equal measures of **gin** and lemon juice with a dash of **Angostura Bitters**, and soda.

tom ka gai (*Thai*) [tum kah gye] Spicy chicken soup with **coconut milk**. (*See also* **tom yam** *below*.)

tom yam (*Thai*) [tum yum] Spicy **clear soup** made with **lemon grass**. (*See also* **tom ka gai** *above*.)

tom yam kung (*Thai*) [tum yum koong] Tom yam with prawns added.

tonic Short for 'Indian tonic water', a non-alcoholic aerated water flavoured with **quinine**, used as a mixer in various **cocktails** and mixed drinks, e.g. **gin** and tonic. (It is called 'Indian' because the beverage was developed in India where quinine was used as a protection against malaria.)

tonno (*Ital.*) Tuna (fish).

torte (*Ger.*) [tor-tuh] Rich cake made in layers, sandwiched together with **cream** and covered with chocolates, fruit or nuts. (*Compare* **tourte**.)

tortellini (*Ital.*) (*pl.*) [tor-tell-een-ee] **Pasta** with a filling of meat or cheese twisted into small rings. (Literally 'tiny cakes'. According to legend, said to resemble Venus's navel.)

tortilla (*Mex.*) [tor-tee-ya] Flat, round **pancake** made of maize (cornmeal) served hot and usually filled with meat or beans and a sauce. (*See also* **enchilada**, **nacho** *and* **taco**.)

toscano [toss-kah-noh] Sourdough bread containing pieces of green olive. (In Italian, *Toscano* means 'Tuscan' or 'from Tuscany'.)

tournedos (*Fr.*) (masc.) [toor-nay-doh] Small round slice of beef **fillet** from the centre of the 'eye'. (Literally 'turn (the) back'.) (*See* **bifteck**, **filet mignon**, **médallions**, **mignonnettes** *and* **noisettes**. *See also below*.)

tournedos Rossini Tournedos served on fried bread, traditionally garnished with **foie gras** and **truffles**. (Devised by Gioacchino Rossini, 1792–1868, composer of operas and famous **gastronome**.)

tourné(e) (*Fr.*) (adj.) [toor-nay] Turned, shaped.

tourte (*Fr.*) (fem.) [toort] Round pie or tart. Savoury *tourtes* have a pastry lid, sweet ones usually do not. (Not to be confused with **torte**.)

traminer One of a group of grape varieties used to make white wine, the most popular of which is **gewürztraminer**.

tranche (*Fr.*) (fem.) [trahnsh] Slice (of fish, meat, etc.); rasher. (*See also* **tronçon** *and* **darne**.)

trancheur/se (*Fr.*) [trahnsh-ur/urse] Person responsible for carving (from a buffet or carving trolley).

trattoria (*Ital.*) Italian restaurant, usually quite small.

trifle **Dessert** dish of sponge cake (often soaked in sherry) topped with fruit, **custard** and cream.

tripe Lining of cow's stomach. (French *tripe* [treep].)

triple sec Very sweet white **curaçao**, e.g. **Cointreau**.

tronc (*Fr.*) [tronk] A box for tips received by restaurant staff and later shared out.

tronçon (*Fr.*) (masc.) [tro(n)so(n)] Slice, usually of large flat fish cut through the bone. (*See also* **darne**.)

trotters Feet of hoofed animal, esp. pig's feet.

truffle (1) Edible fungus (**tuber**) found underground, esp. near the roots of oak trees, highly prized as a savoury delicacy. (French *truffe* (fem.) [troof]; Italian *tartufo*.)
(2) Rich, creamy chocolate **sweet**, shaped into a ball; a chocolate truffle. (So called because it resembles the truffle tuber.)

truite (*Fr.*) (fem.) [trweet] Trout.

truite au bleu (*Fr.*) [trweet oh bluh] Freshly caught trout, cleaned, sprinkled with vinegar and plunged into a **court-bouillon** as a result of which the trout turns blue and curves into a bow. (*See also* **bleu**.)

tuber Vegetable that swells underground, e.g. potato, **Jerusalem artichoke**, **truffle**.

tuile (*Fr.*) (fem.) [tweel] Almond wafer curved like a French tile and served as a **petit four**. (*Tuile* means 'tile'.)

tumbler Simple, straight-sided glass for serving water and soft drinks. (*See* Chapter 13 *and* subsequent diagram of glassware.)

tureen [ture-reen] Deep bowl with a lid from which soup is served at the table. (Not to be confused with **terrine**.) (*See also* Chapter 3.)

turkey buffet (*Austral.*) Joint of turkey breast with the legs and wing tips removed; **smoked** turkey breast.

Turkish coffee Strong, dark, sweet **coffee** served in small cups; **Greek coffee**.

turmeric Mildly peppery spice, bright yellow when ground to powder, used in most **curry** mixtures, usually called *haldi* in India.

tutti frutti Ice cream mixed with **diced glacé** fruit that has been **macerated** in alcohol. (Literally 'all the fruits' in Italian.)

tzatziki (*Grk* & *Turk.*) [tsat-seekee] Yoghurt mixed with **garlic** and cucumber and served as an **appetiser** or dip with the **meze**.

udon Thick, soft, bland **noodle** made of wheat flour, popular in Japan and China. Udon noodles are usually boiled in stock and are often served as a snack. (*Compare* **soba**.)

un/une (*Fr.*) (masc./fem.) [u(n)/oon] A. (The indefinite pronoun, e.g. *un* **goujon** *de* poisson, *une* **quiche**.)

Underberg Proprietary brand of German **bitters**, taken both as a tonic and a **digestif**; it is drunk straight from the bottle in one mouthful.

underfillet Tenderloin or **fillet** of beef.

underliner Plate, usually with a **doily** on it, placed under another dish, such as a soup bowl; an underplate. (*See* Chapter 7.)

vacherin (*Fr.*) (masc.) [vasheri(n)]
(1) Round, flat, soft cow's-milk cheese. (*Vache* means 'cow'.)
(2) Dessert consisting of layers of crisp **meringue** sandwiched with **cream**, similar to **pavlova**.

vacuum infused coffee *See* **Cona coffee**.

Valrhona Brand of **couverture** chocolate; high-quality brand of chocolate used for cooking.

vanilla Sweet and fragrant flavouring obtained from the bean of the vanilla orchid (*Vanilla planifolia*). (*See also* below.)

vanilla bean Long pod of the **vanilla** orchid (*Vanilla planifolia*) containing the seeds producing the distinctive flavour. Dishes flavoured with a vanilla bean might have a slightly speckled appearance resembling dust.

vanilla essence Concentrated extract of the **vanilla bean** used to flavour sweet dishes in place of the bean.

vanilla sugar Sugar stored in a container with a **vanilla bean**, which imparts the vanilla flavour to the sugar.

varietal Adjective used to describe wine made from a particular variety of grape, e.g. **chardonnay**. (*See also* **generic**.)

veal Calf's meat. The finest comes from calves fed exclusively on milk. (French **veau**, Italian **vitello**.) (*See also* **blanquette**, **girella**, **osso buco**, **paillard**, **saltimbocca** *and* **Wiener schnitzel**.)

veau (*Fr.*) (masc.) [voh] Veal.

vegan [vee-gan] Strict **vegetarian**; person who eats no animal products at all, not even milk or eggs.

vegetarian Person who doesn't eat animal (or fish) products, esp. the flesh of slaughtered animals. Many vegetarians do eat eggs and consume dairy products (milk, cheese, etc.). (*Compare* **vegan**.)

velouté (*Fr.*) (masc.) [vuh-loot-ay] Thick velvety sauce made from a light **stock**, rather than milk, and a white **roux** to which **cream** is often added.

venaison (*Fr.*) (fem.) [venay-so(n)] Venison; the meat of any large **game** animal. (*See also* **chevreuil**, **poivrade**, **salmis** *and* below.)

venaison, sauce (*Fr.*) [sohs venay-so(n)] **Sauce poivrade** to which fresh cream and redcurrant jelly have been added, traditionally served with **game** and dishes '*en* **chevreuil**'. (*See also* **venaison**.)

venison [ven-sun] Meat of deer. (*See also* **venaison**.)

verde (*Ital.*) [vairduh] Green. (*See also* **salsa verde**.)

verjuice [vair-juice] Unfermented grape juice; juice of unripe grapes. (French *verjus* [vair-zjoo], literally 'green juice'.)

vermicelli (*Ital.*) (*pl.*) [vairmi-chelly] Long thin threads of **pasta** sometimes coiled into rings; **noodles**. (Literally 'small worms'.)

vermouth [ver-muth] Flavoured **fortified wine** available in three principal styles: dry (or **French**), sweet (**rosso** or Italian) and **bianco**, which is golden and medium-sweet. (*Rosso* literally means 'red' and *bianco* 'white'.)

viande (*Fr.*) (fem.) [vee-ond] Meat.

vichyssoise (*Fr.*) (masc.) [vee-shee-swaz] Creamy soup made of **puréed** leeks and potatoes, served chilled and garnished with **chives**. (*See also* **garbure** *and* **potage**.)

Vienna coffee *See* Chapter 16.

Vietnamese mint *See* **mint, Vietnamese**.

vin (*Fr.*) (masc.) [vi(n)] Wine.

vinaigrette (*Fr.*) (fem.) [vi(n)-ay-gret] Mixture of oil and **vinegar** or lemon juice used to dress salads. Other ingredients, esp. **mustard** or herbs, may be added. (*See* **French dressing** *and* **wine vinegar**.)

vindaloo (*Ind.*) [vin-daloo] Very hot S Indian **curry**, spiced and flavoured with **vinegar**.

vinegar Fermented acidic **liquor** made from wine, cider, **malt**, **spirits** or rice. (French *vinaigre;* literally 'sour wine'; Italian *aceto*.) (*See* **balsamic, malt, rice** *and* **wine vinegar**; *see also* **vinaigrette**.)

vin ordinaire (*Fr.*) [vi(n) or-din-air] Moderately priced everyday wine; **house wine** bought in bulk by an **establishment** and often served from **carafes**.

vintage [vin-tij] (1) (noun) Season when grapes are harvested (as in *the vintage was late this year*).
(2) (noun) Wine made from the season's produce; the year the wine is made (as in *the 1999 vintage*).
(3) (noun) Wine from a single year (as in *1999 was a good vintage*).
(4) (adj.) Wine or **port** of special quality (*vintage wine*); of high quality (as in *a vintage year*).

virgin Without alcohol. (*See* **cocktail** *and* Chapter 14.)

virgin olive oil Finest olive oil; oil extracted from the first pressing of the olives, and which contains no additives.

vitamin Any of a group of substances found in food that help regulate body processes essential to health and growth. Vitamins are classified by letters and numbers (vitamin A, B_2, C, D_3, E, etc.).

vitello (*Ital.*) **Veal**.

vitello tonnato (*Ital.*) [vitello ton-ah-to] **Veal** with tuna sauce.

vodka Spirit made from grain; in the West, usually, colourless and flavourless, but in Russia and Poland often flavoured. It is most popular as an ingredient in **cocktails**, e.g. a **Bloody Mary**.

voiture (*Fr.*) [vwah-tyoor] Trolley used in the dining-room for various purposes such as carving or carrying hors d'oeuvres, sweets, cheeses or liqueurs. (Literally 'conveyance' or 'car'.)

volaille (*Fr.*) (fem.) [voll-eye-yuh] Poultry.

vol-au-vent (*Fr.*) (masc.) (*pl.* **vols-au-vent**) [voll oh vah(n)] Round case of **puff pastry** filled with a savoury mixture held in a creamy sauce and capped with a lid. Tiny ones are served as cocktail **savouries** and larger ones as an **entrée**. (Literally 'flight in the wind'.) (*See also* **croustade**.)

VSOP Very Superior Old Pale, a technically meaningless term used on labels to market **cognac**.

waffle [wofful] Small crisp cake made of **batter**, baked in a waffle iron, which gives it its shape and marks it with a distinctive pattern. (From Dutch *wafel*, meaning 'wafer'.) (*Compare* **jaffle**.)

waiter's friend Tool that is a combination of bottle-opener and corkscrew. (*See* Chapters 13 *and* 15.)

Waldorf salad Salad made with lettuce, **diced** apples, walnuts, celery and **mayonnaise**. (Invented at the Waldorf Hotel, New York.)

warrigal greens Green-leaf vegetable (*Tetragonia tetragonoides*) used in salads, or cooked and used as spinach; New Zealand spinach.

wasabi (*Jap.*) [woss-ah-bee] Hot, pungent Japanese **condiment** made into a pale green paste from a vegetable that resembles **horseradish**. It is served with **sashimi** and **sushi** and added to soups. Wasabi is available as a powder or ready-mixed paste.

watercress Green leaf vegetable (*Nasturtium officinale*) with small, round leaves and a pungent flavour. (*See also* **cress**.)

water ice Iced dessert made of **puréed** fruit, **sugar syrup** and egg white. (*See* **sorbet**).

wattle seed (*Austral.*) Milled seed of the wattle (*Acacia*) species. Wattle seed is used to flavour both biscuits and bread, and many sweet dishes, esp. cream desserts.

wedges Chunks of vegetable, usually potato, fried or baked and served with **soured cream** and sweet **chilli** sauce.

Wellington *See* **beef Wellington**.

Welsh rabbit/Welsh rarebit (*UK*) **Savoury** made of melted cheese mixed with milk, seasonings, etc. on hot toast.

whisky/whiskey Spirit **distilled** from **fermented** corn (barley, rye, wheat or maize). There are many styles. **Scotch** whiskies are the most popular. Most Scotch whiskies are blends of **malt whiskies** (made from barley) and grain whiskies (distilled in bulk from fermented wheat and maize). American and Irish whiskey are spelt with an 'e'. Other varieties drop the 'e'. (*See also* **bourbon** *and* **rye**.)

whitebait Sprats; tiny silver fish, usually fried and eaten whole.

white pepper *See* **pepper** *and* **peppercorn**.

wholegrain Made with complete grains (e.g. bread) or whole seeds (mustard). Wholegrain bread is also known as fullgrain or multigrain bread.

Wiener schnitzel [veener shnit-sell] **Veal escalope** or **cutlet** coated in fine breadcrumbs and **sautéed** or deep-fried. Traditionally garnished with lemon slices and **anchovies**. (From Wien (Vienna), capital of Austria.) (*See* **schnitzel**.)

wild rice Seeds of a cereal, *Zizania aquatica*, native to N America. Brown when raw, it turns slightly purple when cooked and has a nutty taste and slightly chewy texture. Not related to common white rice.

wine (1) Alcoholic drink made from fermented grape juice, e.g. **burgundy**, **claret**, **champagne**.
(2) Any **fermented** vegetable juice that is **distilled** to make spirits.

wine vinegar **Vinegar** distilled from (grape) **wine** and used in **béarnaise sauce**, **vinaigrette**, etc. Some are infused with herbs, spice or fruit, e.g. **tarragon**, **chilli** or **raspberry**. (*See* **balsamic vinegar**, **French dressing** *and* **malt vinegar**.)

witchetty grub Long fat white grub found on wattle roots; usually eaten roasted. Witchetty grub soup is available canned.

witloof/witlof Chicory; white or Belgian **endive** (*Cichorium intybus*). (Flemish for 'white leaf'.)

won ton (*Chin.*) Small **dumpling**; savoury mix, esp. minced pork, enclosed in a won ton wrapper (made of egg noodle dough). They are steamed or fried or added to soup. (*See also* **dim sim/sum**.)

Worcestershire/Worcester sauce [wuster(shuh)] Spicy sauce made of **vinegar**, molasses, sugar, spices, etc. (Worcestershire is a county in W England.)

wrap 'Sandwich' made of **pita, roti, mountain bread** or similar flat bread wrapped around a filling. Wraps are served rolled up in a tube shape.

wurst (*Ger.*) [voorst] Sausage.

yabby/yabbie (*Austral.*) Small freshwater **crayfish** (*Cherax destructor*) with large pincers so that it looks like a miniature **lobster**. (*See also* **marron**.)

yakitori (*Jap.*) Chicken kebab grilled while being basted in a soy sauce, rice wine and sugar sauce. (*Yaki* means 'grill' and *tori* means 'bird'.)

Yorkshire pudding (*UK*) Savoury **batter pudding** traditionally served with roast beef.

yum cha (*Chin.*) Banquet usually served mid-morning and lasting over three hours. Many small dishes, or **dim sim**, are served on a central turntable from which guests help themselves. (*Yum* means 'drink', *cha* 'tea'.) (*See also* **won ton**.)

zabaglione (*Ital.*) [zab-ay-on-nay] Light foamy dessert made by whipping egg yolk, sugar and alcohol (traditionally **marsala**) over a low heat. Served tepid. (*See* **sabayon**.)

zakouski/zakuski (*Russ.*) [za-koo-skee] (1) **Appetisers**, e.g. tiny open sandwiches spread with **caviare**, smoked sausage, etc. usually served with **vodka**.
(2) Small sweet items served after a meal. (Literally 'little bites'.)

zest (of **citrus** fruit) Outer rind of **citrus** peel without the pith, and the scented oil it produces. It is used to enhance a citrus flavour.

zingara (*Fr.*) (adj.) [zi(n)-gahra] Containing **paprika** and tomato. (*Zingaro* means 'gypsy'.)

zucchini Variety of vegetable marrow or squash (*Cucurbita pepo*) also known as **courgette**. The flowers can be stuffed with a **farce** and deep-fried.

zuppa (*Ital.*) [zuppah] Thick soup. (*See also* **brodo** *and* **minestra**.)

zuppa inglese (*Ital.*) [zuppa inglay-say] **Dessert** or **pudding** made from sponge cake, chocolate and **custard**; **trifle**. (Literally 'English soup'.)

zwieback (*Ger.*) [zvee-back] (1) Hard, crisp biscuit or rusk.
(2) Toasted thin slice of sweet cake. (Literally 'twice baked'.)

Index